Great Observatories of the World

A FIREFLY BOOK

Published by Firefly Books Ltd. 2005

Originally published as *Les Grands Observatories du Monde*
ISBN: 2-04-760026-X
Copyright © 2002 Bordas/HER 2002
English translation copyright © 2005 Firefly Books

Publisher Cataloging-in-Publication Data (U.S.)

Brunier, Serge.

Great Observatories of the World/Serge Brunier and Anne-Marie Lagrange.

Originally published as Les Grands Observatoires du Monde ; Bordas, HER : France, 2002.

[240] p. : chiefly col. photos., ; cm.

Includes bibliographical references and index.

Summary: An in-depth description of observatories on Earth and in space, including telescopes planned for the future, history of astronomy and telescopes and essays on technology involved in observatories.

ISBN 1-55407-055-4

1. Astronomical observatories. I. Lagrange, Anne-Marie. II. Title.

522.1   22   QB81.B78   2005

Library and Archives Canada Cataloguing in Publication

Brunier, Serge
Great observatories of the world/Serge Brunier and Anne-Marie Lagrange ; foreword by Catherine Césarsky.

Translation of: Les grands observatoires du monde.

ISBN 1-55407-055-4

1. Astronomical observatories. I. Lagrange, Anne-Marie II. Title.

QB81.B7813 2005   522'.1   C2004-907015-0

Published in the United States in 2005 by
Firefly Books (U.S.) Inc.
P.O. Box 1338, Ellicott Station
Buffalo, New York 14205

Published in Canada in 2005 by
Firefly Books Ltd.
66 Leek Crescent
Richmond Hill, Ontario L4B 1H1

Printed in Italy

Acknowledgments

Serge Brunier thanks the many astronomers, passionate researchers all, who helped him make this book a reality. Whether at brief encounters on the way to the observatory, during coffee breaks at international symposia, or simply through enriching e-mail exchanges, all have contributed. These individuals include: Danielle Alloin, Jean-Luc Beuzil, Lain Blanchard, Jérôme Bouvier, Fabienne Casoli, Fred Chaffee, Jean-Charles Cuillandre, Mark Dickinson, Philippe Dierickx, Daniel Enard, Olivier Le Fèvre, Wendy Freedman, Robert Fugate, Roberto Gilmozzi, Ronald Gilliland, Stéphane Guilloteau, François Hammer, Jean-Paul Kneib, Antoine Labeyrie, Pierre Léna, Jean-Pierre Luminet, Barry Madore, David Malin, Michel Mayor, Yannick Mellier, Guy Monnet, Francesco Paresce, Bryan Penprase, Hubert Reeves, Marc Sarazin, Sylvie Vauclair, Alfred Vidal-Madjar, Christian Veillet, Richard West, Ray Wilson, David Wittman and all their colleagues…

Anne-Marie Lagrange thanks Pascal Rubini and his colleagues David Mouillet and Gilles Duvert for their readings of the manuscript and their always pertinent comments, as well as her children, Marion and Étienne, for their patience.

Edited by: Sylvie Cattaneo-Naves, Laurence Alvado, Sandrine Vincent, in collaboration with Érick Seinandre

Copy editors: Madeleine Biaujeaud, Jean Delaite

Index: Érick Seinandre

Artwork: Emmanuel Chaspoul

Graphics and layout: Gilles Seegmuller, Michel Delporte

Cover: Véronique Laporte

Diagrams and tables: Laurent Blondel

Iconography: Carine Lepied, Christine Varin

Production: Nicolas Perrier

English translation: Klaus R. Brasch

Great Observatories of the World

FIREFLY BOOKS

Great Observatories of the World

Serge Brunier
Anne-Marie Lagrange

Foreword by Catherine Césarsky
Director General of the European Southern Observatory (ESO)

This book is a collaborative effort between an award-winning science journalist and a widely recognized professional astrophysicist, and provides a unique overview of both contemporary world observatories and of telescopes in the future.

Thanks to remarkable advances in optics and electronics, astronomers have made impressive new discoveries in recent decades. Moreover, they no longer have to spend long, uncomfortable nights in the dead of winter in observing cages on top of their telescopes, or with their eyes glued to a dimly lit eyepiece reticle following a guide star. During the 1970s, they began observing in the comfort of heated rooms, and controlling their telescopes, support instrumentation and images with computers.

In due course, electronic detectors with far greater sensitivity and better contrast replaced the photographic plate. Radio astronomy also made impressive advances, including pioneering the use of aperture synthesis and interferometry, which greatly increased image resolution and quality. Space-based astronomy was the next great stride forward; by rising above the Earth's atmosphere, astronomers were no longer limited to the visible end of the spectrum, but now observed at wavelengths that included ultraviolet, X-ray and gamma-ray emissions. This helped reveal a far more violent cosmos than hitherto imagined, by showing a universe filled with incredible explosions and energy, and with black holes devouring everything in their reach.

Sadly, one of the great ironies of the 20th century is that while astronomers can now probe the universe at all possible wavelengths, from Earth as well as from space, ever-increasing light pollution has progressively rendered the night sky less and less accessible to the average citizen.

In their quest for ever-sharper celestial images, Earth-based astronomers and observatories have long been hampered by such factors as instrument flexure, wind and atmospheric turbulence. To overcome this, they have developed flexible, computer-controlled optics that compensate for and correct such problems automatically. In fact, with the advent of active and adaptive optics, ground-based telescopes can now compete effectively with much costlier space-based instruments like the Hubble Space Telescope. As we begin the 21st century, researchers have fully integrated ground and space-based instrumentation to maximize their effectiveness and to plan future projects. Collectively, this is leading to the establishment of a "virtual observatory," a huge repository of data and information accessible to the worldwide astronomical community, as well as to members of the general public who are interested in the findings of the great observatories.

In the United States, foundations and private donations have helped finance the most ambitious ground-based telescopes (including the twin Keck instruments in Hawaii), while in space, the most important projects have been funded by NASA. In Europe, intergovernmental collaborations among nations have helped make the largest projects possible. In space, although the European Space Agency (ESA) cannot match NASA's budgets, it has launched several successful missions, including Giotto, the first probe to a comet; Hipparcus, the first astrometric space observatory; and ISO, the first infrared space telescope. The latter also played a major role in collaboration with the SOHO solar mission. With establishment of the ESO at Cerro Paranal and a remarkable degree of collaboration, Europe has taken the lead in ground-based astronomy. Shared expertise among several national observatories led to the interferometers at Calern and Cambridge, research on exoplanets at Haute-Provence and La Silla observatories, use of artificial stars for adaptive optics at Calar Alto, and so on. After their success with active and adaptive optics through the New Technology Telescope (NTT) and Adonis, as well as the deployment of the Very Large Telescope (VLT) and the Very Large Telescope Interferometer (VLTI), European astronomers are proposing the construction of an Overwhelmingly Large Telescope: the 100 m OWL.

In November 2003, a groundbreaking ceremony was held for a joint European and North American project: ALMA, a giant millimeter and submillimeter interferometer located in Chile at an elevation of 5,100 m on a vast plateau in the Atacama Desert. This project, which includes a number of other international partners, is the first example of scientific collaboration of this nature. It is perhaps fitting that astronomers, who are attempting to decipher the whole universe, should be the first to undertake a project of such truly global proportions.

However, our readers can rest assured that despite the daunting technological advances described in this book, professional astronomers are just as overwhelmed as they are by the great beauty and mysteries of the cosmos.

Table of Contents

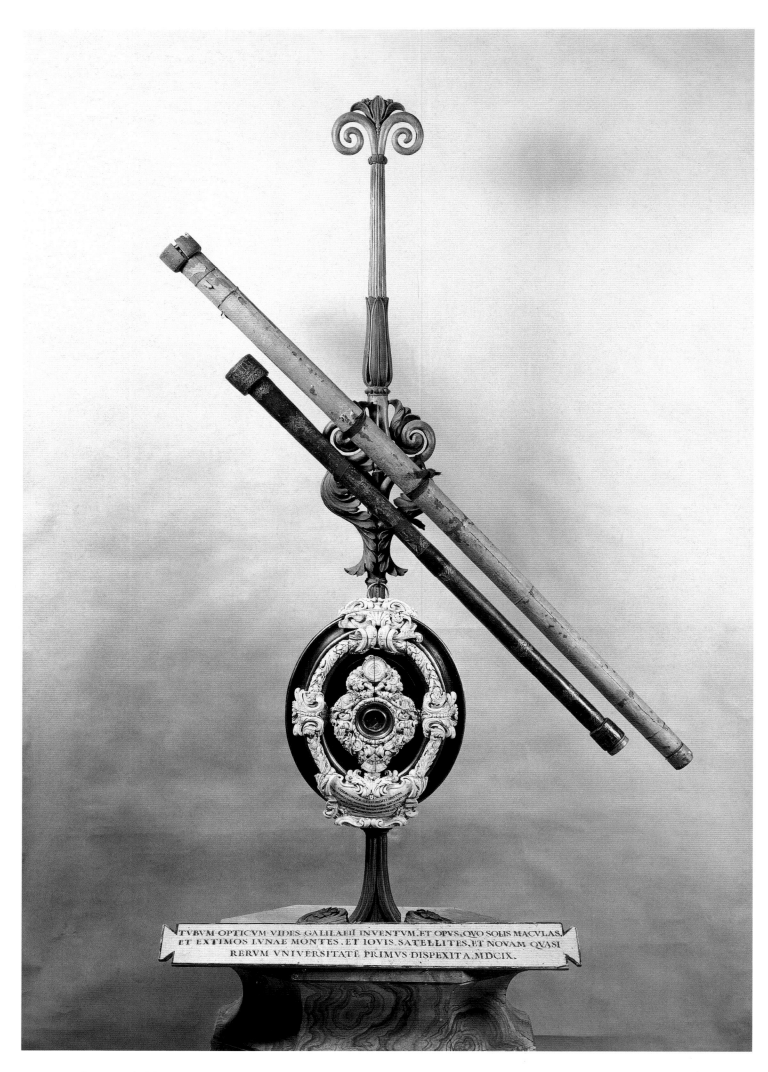

■ *Humans have probably contemplated the sky since their earliest days on Earth. However, the ancient science of astronomy was revolutionized in November 1609, when Galileo first pointed a telescope at the Moon, the planets and the stars. Galileo's first two telescopes are housed at the Museum of the History of Science in Florence.*

From Galileo's refractor to the Mount Palomar reflector

On the evening of November 30, 1609, in the powerful Republic of Venice, Galileo (Galileo Galilei, 1564–1642) hurried across San Marco's square in the cold night air. In their hurry to get home, passersby did not notice that the skittish Tuscan savant carried a long wooden tube under his mantle. Passing the Basilica of San Marco to his right, Galileo crossed the square and turned toward the Mole. There, out of sailors' sight, he set up his groundbreaking spyglass overlooking the Venetian lagoon. Several months before, he had tested his telescope from the top of the campanile, showing the Doge images of distant ships on the horizon some two hours before the port lookout spotted them.

As he gazed through the ocular of his 20-power telescope this evening, Galileo's target was the four-day-old crescent Moon shimmering in the waters of the Grand Canal…

While never claiming to have invented the astronomical telescope, which was probably done by Dutch spectacle-maker Hans Lippershey (1570–1619), Galileo crafted his with great care. He ground and polished the concave and convex lenses he needed for adequate magnification. This likeable physicist, a peer of Newton and Einstein, was also the first modern astronomer. Observing night after night from November 30, 1609, through March 2, 1610, he not only revealed the immensity of the Universe but was secure in the knowledge that the Earth is not at its center. With magnifications of 20 or 30, lunar features appeared in stark relief to Galileo and were clearly comparable to similar terrestrial features. The Moon was not after all the ethereal sphere envisaged by the philosopher Aristotle, but a separate *world* with craters, mountains and plains…. Though contrary to church doctrine at the time, Galileo's discovery that Jupiter was orbited by its own set of moons directly supported Copernicus' heliocentric theory of the Solar System. When Galileo pointed his telescope at the darkest areas of the sky, where nothing was visible to the naked eye, he saw stars never seen before then. More astounding still, when this first telescope was pointed at the ghostly Milky Way, it resolved multitudes of faint, tightly packed stars. Apart from revealing the immensity of space, Galileo's telescope also shattered the ancient Greek doctrine that the cosmos is closed and composed of interlocked crystalline spheres.

As extensive contributions by amateur astronomers have demonstrated since the 18th century, the marvels of astronomy are accessible to all. You don't need complex instrumentation, great wisdom or high social standing to verify for yourself Galileo's extraordinary discoveries. During the 17th and 18th centuries, dauntingly long telescopes became popular throughout Europe and artists, nobility and ordinary people alike, overwhelmed by the vastness of space, speculated about other planets in this seemingly infinite universe.

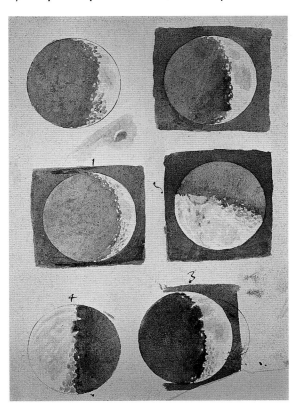

■ *Drawings of the Moon by Galileo, as seen through his telescope. A new world was revealed, complete with plains, valleys, mountains and craters.*

■ *Johannes Hevelius at his observatory in Danzig. He compiled the* Firmamentum, *one of the first great atlases of the sky, showing 1,564 stars.*

THE FIRST IMAGES OF THE COSMOS

The Lippershey-type refractor used by Galileo was basically assembled by trial and error, and so had many imperfections. However, by the end of the 17th century astronomers and opticians sought new formulas and designs that might provide bigger, brighter and, above all, sharper images. While capable of showing the phases of Venus, sunspots and lunar craters, Galileo's refractor had an objective lens only 30 mm in diameter, with many optical aberrations. This was not good enough to reveal fainter markings on Jupiter and Mars or the rings of Saturn. Its resolving power was less than 10 arcsec, and it probably did not show stars fainter than 9th magnitude. The naked eye in comparison can resolve details of about 60 arcsec and detect stars to about 6th-magnitude.

Ultimately, even when ground and polished with great care and precision, the lenses used by Galileo and his successors could never have provided really good images. All astronomical refractors at the time and for another century and a half thereafter suffered from one major defect, chromatic aberration. This is because light rays are deflected when entering transparent media like glass; they are refracted or bent from their angle of incidence in proportion to their respective wavelength. In other words, several images of the same stellar object are formed, not only at slightly different focal points but also of different colors!

Most observers had abandoned the Galilean refractor by 1640 in favor of those designed by the great astronomer Johannes Kepler (1571–1630), who improved optical performance by using convex lenses both as objectives and eyepieces. It also soon became apparent that aberrations could be diminished and image quality improved by lessening the surface curvatures of lenses and thus increasing the focal length of telescopes relative to their aperture. This led to construction of some truly gigantic refractors by the middle of the 18th century. With their immense bamboo tubes, huge wooden mounts, and ropes and pulleys to keep them aligned, these instruments required many assistants to operate. They also demanded great dedication on the part of the observer, who had to work under very uncomfortable conditions.

In 1655, Dutch astronomer Christian Huygens (1629–1695) discovered that "Saturn possesses a thin flat ring that is at no point attached to the planet itself." That same year he also discovered Titan, Saturn's largest satellite. In 1659, using a 37 m long refractor, he was the first to detect surface detail on Mars. In addition to the northern polar cap, Huygens recorded a dusky, triangular-shaped structure in the center of the disk. This is Syrtis Major, the red planet's most prominent dark feature. In 1666, Huygens, now a resident of Paris, was nominated to the French Royal Academy of Sciences.

The Paris Observatory, which was founded by Colbert under Louis XIV in 1667, provided Christian Huygens and Jean Dominique Cassini (1625–1712) almost sole access to an essentially

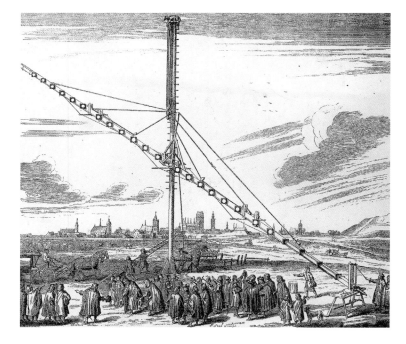

■ *The largest telescope built by Hevelius was 46 m long. All instruments used by this Polish astronomer were described in his* Machina coelestis, *published in 1673.*

unexplored sky. By timing the transit of Syrtis Major across the center of the tiny Martian disk on several successive nights, Cassini was able to calculate the planet's period of rotation and found that it was almost identical to that of Earth.

In 1667, this first director of the Paris Observatory calculated Jupiter's period of rotation. He also discovered the Great Red Spot, a gigantic cyclone still regularly observed on Jupiter today. The Paris Observatory refractor was of exceptional quality, and probably resolved detail to 2 arcsec, and stars to 11.5–12th magnitude. Cassini studied Saturn intensively between 1671 and 1684, and

■ *The Paris Observatory, founded in 1667 under Louis XIV, and the Tower of Marly, which supported the "long refractors that frightened people" used by Cassini, Huygens and Römer.*

discovered four additional satellites: Iapetus, Rhea, Dione and Thethis. In 1675, he noticed that the rings were not uniform but separated into zones of slightly different intensity by a division that was later named after him. Cassini was also the first to understand the physical nature of Saturn's rings, composed he said "of an association of very small satellites of varying motion that cannot be visually differentiated."

Still at Paris Observatory, Danish astronomer Ole Römer (1644–1710) carefully studied the motions of the four Galilean satellites: Io, Europa, Ganymede and Callisto. Unexpectedly, he noted that the eclipses of these Jovian satellites were either "delayed" or "advanced" relative to their predicted times. Römer soon realized that this was due to changes in the distance between the Earth and Jupiter during the year, and that light did not travel infinitely fast, but took longer to reach us when Jupiter was furthest. He estimated the speed of light at about 212,000 km/s, remarkably close to its actual value of 299,792.458 km/s.

During this same period, Johannes Hevelius (1611–1687) in Danzig compiled a magnificent atlas of the Moon, which he published in 1647 in his *Selenographia*. He also continued to build ever-larger refractors, culminating with his famous "celestial machine," a cumbersome 46 m long instrument that never served him well. In fact it became quite obvious that despite the discoveries of Huygens, Cassini, Hevelius and several others including Nicola Fabri de Peiresc (1580–1637) in France, the giant refractors of the time had no real future. They were of inadequate optical quality and clearly too cumbersome to operate.

A new era began in 1672 when the English physicist Isaac Newton (1642–1727) presented the first reflecting telescope to members of the Royal Society of London. This revolutionary design used a concave, highly polished mirror of metal alloy in place of the refractor's objective lens. The mirror reflected and focused light rays back into the optical path, where a flat secondary mirror deflected them to the ocular through a hole in the tube. This very compact telescope yielded sharp and completely color-free astronomical images. With its 37 mm diameter mirror and 160 mm focal length, it was less than .3 m long; a refractor of similar aperture was typically 3 to 10 m long!

■ *Isaac Newton presented the first reflecting telescope to the Royal Society of London in 1672. This instrument was equipped with a 37 mm diameter mirror and had a magnifying power of 38.*

■ *Toward the end of the 18th century, it became fashionable for aristocrats to buy and observe with telescopes produced by optical craftsmen.*

THE OPTICIANS ENTER THE SCENE

Although clearly innovative, the Newtonian reflector, which had also been suggested by the likes of Galileo, René Descartes (1596–1650), James Gregory (1638–1675) and Laurent Cassegrain (1628–1693), took over 200 years to become the dominant design. While superior in optical quality, polished metal mirrors had some major drawbacks, including poor reflectivity and a tendency to tarnish rapidly. Because of this, most astronomers avoided them. However it took a century and a half after Galileo before the "curse" of chromatic aberration was finally overcome, by English optician John Dollond (1706–1761), who designed the first "achromatic" refractor in 1758. After much trial and error, Dollond discovered that chromatic aberration could be almost completely eliminated by combining two glass elements in the same objective, one convex and the other concave, and each ground of glass with a different refractive index. The era of the refractor was about to begin.

Most achromatic refractors built at the end of the 18th century were called "transit" telescopes. They were fixed in the horizontal plane, and could only move vertically in a north/south direction. Such telescopes were used primarily for sky surveys. People were keenly interested in interpreting the "great cosmic timepiece" to establish a reliable system of coordinates for navigators and explorers. Charles Messier (1730–1817), astronomer to Louis XV, did not compile his catalog of nebulous astronomical objects to find out more about them, but to avoid confusing them with comets. The highly elliptical orbits of comets helped clarify the "celestial mechanics" of the Solar System and the universal laws of gravity that Newton had recently formulated.

Transit telescopes helped refine positional measurements and also generate increasingly accurate star charts. John Flamsteed (1646–1719), Astronomer Royal and founder of Greenwich Observatory, was able to catalog only 2,866 stars to an accuracy of about 10 arcsec. This was quickly surpassed by the work of Joseph Jérôme Lefrançois de Lalande (1732–1807), who cataloged 47,390 stars.

In the 17th century Galileo and his successors surveyed the Solar System, but in the 18th century astronomers finally turned their attention to the stars. They devoted much care to celestial cartography because they now realized that stars are not in fact "fixed" as believed since antiquity, but actually move across the sky, albeit very slowly.

Thus in 1718, the famed Edmond Halley (1656–1742) compared the position of Aldebaran in Taurus as recorded by Athenian astronomers in the year 509 during a lunar occultation, with its position in 1718. He found that in the intervening 1,209 years this brilliant red star was displaced 8 arcmin farther south; the equivalent of one quarter the apparent diameter of the Moon! By the end of the 18th century, Italian astronomer Giuseppe Piazzi (1746–1826), the discoverer of Ceres, the first asteroid, showed that many stars exhibited clear proper motion. The most spectacular example was 61 Cygni, a beautiful red binary star moving across the sky at more than 4 arcsec per year.

■ *At the Royal Greenwich Observatory, built during the reign of Charles II, John Flamsteed put the great .32 m equatorial refractor into service in 1859.*

In 1806, the German optician Joseph Fraunhofer (1787–1826) built the first spectroscope, which is probably the most important analytical instrument of modern astronomy. When he analyzed the Sun and the stars with this instrument, Fraunhofer discovered that each had unique spectral signatures, defined by emission and absorption lines. He also used spectroscopy to better understand the optical properties of different glass lenses, leading to the construction of astronomical refractors of exceptional quality.

After Piazzi, Friedrich Bessel (1784–1846) started observing the intriguing 61 Cygni with the .16 m Fraunhofer refractor at Königsberg Observatory. He measured the binary star's position against background stars with great precision and found a parallax angle of about 0.3 arcsec. Bessel was able to calculate the distance to a star by simple triangulation for the first time. He found that 61 Cygni was over 10 light-years from Earth, or more than 100 trillion km. By this single stroke, our insight into the dimensions of the cosmos expanded enormously.

■ *Lord Rosse observed the "spiral nebula" (M51) in Canes Venatici in detail for the first time in 1850.*

THE FIRST GIANT TELESCOPES

It was a devoted organist, William Herschel (1738–1822), who stopped looking at the stars as mere points of light or celestial coordinates and inquired into their true physical nature. Herschel was also an excellent optician and built reflecting telescopes of exceptional quality. On March 13, 1781, while cataloging stars in the constellation Gemini with a .23 m reflector of 3 m focal length, he accidentally discovered the planet Uranus and two of its satellites, Titania and Oberon. After being appointed Astronomer Royal by King George III of England, Herschel continued to build ever-larger telescopes, culminating in 1789 with the incredible "forty-foot giant." With its 1.2 m mirror and 12 m focal length, this was far and away the largest telescope in the world at the time. In an unsuccessful effort to measure stellar parallaxes as distance indicators, Herschel discovered the first true double stars. In 1803 he wrote "I have amassed a series of observations of double stars over the past 25 years which, if I am not mistaken, prove that many stars are not simply apparent doubles but are true binary associations, intimately linked through mutual forces of attraction."

In 1800, William Herschel made another fundamental discovery in physics and astrophysics. He dispersed the Sun's rays with a prism and measured the temperature of each primary color with a thermometer. Noticing that the temperature rose at the extreme red end of the spectrum, he continued beyond that point. Since the temperature continued to rise, Herschel realized there were rays beyond the visible end of the spectrum; he had just discovered infrared light.

After Herschel, William Parsons, third earl of Rosse (1800–1867), continued to build giant telescopes in England, culminating in 1845 with the "Leviathan of Parsonstown" at Birr Castle. This four-metric-ton telescope, equipped with a 1.8 m diameter metal mirror, was extremely difficult to operate. While not extensively used, this telescope led Lord Rosse to the discovery of the "spiral nebula" (M51) in Canes Venatici.

■ *Lord Rosse's 1.8 m diameter telescope was 15 m long and weighed four metric tons. It was used until 1870.*

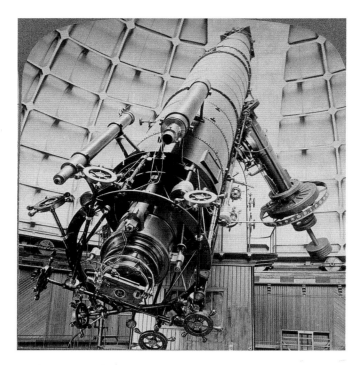

■ *The .89 m refractor at Lick Observatory, Mount Hamilton, is shown above. Until the end of the 19th century most astronomical observing was done visually at the eyepiece of manually controlled instruments like this.*

Apart from occasional efforts mainly by amateurs like Herschel and Lord Rosse, large reflecting telescopes did not really see widespread use until the 20th century. Professional astronomers favored the near-perfect optics of refractors, and as their horizons expanded in the 19th century, the need for precision and higher resolution also increased. This led to a veritable race between European and American astronomers and opticians to build the "largest astronomical refractor in the world." This competition was primarily driven by research on double stars. By accurately measuring the separation and orbital parameters of binary stars, it was possible to determine both their masses and intrinsic luminosities. The first modern, "largest" refractor in the world was built by Fraunhofer and installed by Wilhelm Struve (1793–1864) at Dorpat in Russia. This .24 m telescope had a focal length of 4.3 m. Again in Russia a few years later, Tsar Nicolas I commissioned the most powerful astronomical telescope of his time, a superb .38 m equatorial refractor of 6.88 m focal length. Completed in 1886 at Pulkovo Observatory on a hill near St. Petersburg, this famous Russian observatory later acquired an even larger refractor with a .76 m objective lens.

The exploits of Russian astronomers were viewed with a mixture of admiration and envy in the rest of Europe and particularly in the United States, where no significant astronomical facility existed at the time, prompting both public and private sources to fund construction of ever-larger telescopes. Between 1873 and 1897, no less than a dozen giant equatorial refractors were installed at Berlin and Potsdam in Germany; in Vienna, Austria; at Greenwich and Herstmonceux in England; and at Nice and Meudon in France. Likewise in the United States, major observatories were established in Charlottesville, Washington, Pittsburgh, Lick and Yerkes.

The Americans ultimately built the two largest refractors in the world. An .89 m refractor of 17 m focal length was put into service in 1888 at Lick Observatory on Mount Hamilton, California, and the pinnacle of giant refractors was reached in 1897 at Yerkes Observatory near Chicago. Still in service today, this famed instrument boasted a 1 m objective lens and a focal length of 19 m. These telescopes were especially suited for visual studies of the planets and their surroundings. This era began in 1877 when Asaph Hall (1829–1907) discovered the two Martian moons, Phobos and Deimos, using the .66 m refractor at the U.S. Naval Observatory in Washington. Spectroscopy and photography were also coming into play at this time, with the first photos of the Sun and the Moon taken in middle of the 19th century. In July 1850, William Bond (1789–1859) and John Whipple (1822–1870) took the first photograph of a star, Vega, with the .38 m refractor at Cambridge

■ *The great .76 m refractor at Pulkovo Observatory near St. Petersburg was put into service in 1886.*

Observatory in Massachusetts. Thirty years later, on September 30, 1880, Henry Draper (1837–1882) photographed the Orion Nebula with a .28 m refractor made by the celebrated American optician Alvan Clark (1804–1887). By 1887, advances in photography were so spectacular that the director of the Paris Observatory, Ernest Mouchez (1821–1892), persuaded the international scientific community to cooperate on a "photographic atlas of sky." Eighteen identical .33 m refractors of 3.4 m focal length were manufactured for this project by opticians Paul Henry (1848–1905) and Prosper Henry (1849–1903) and installed at various latitudes around the world.

Undoubtedly the great refractors' most important contributions were in the field of double-star astronomy. Alvan Clark, creator of most of the big objective lenses of that era, discovered the companion star to Sirius in 1862, while testing the new .48 m telescope at Dearborn Observatory near Chicago. Though predicted by Freidrich Bessel in 1842, this was the first white dwarf observed. In 1896, John M. Schaeberle (1853–1924), using the great Lick refractor, discovered a second white dwarf, this time a companion to Procyon.

Despite their undeniable successes, it became quite apparent by the end of the 19th century that refractors were not the wave of the future. Along with the increasing need for more light-gathering power for the new techniques of spectroscopy and photography, it also became clear that the labor involved in casting, polishing and mounting the heavy lenses and 20 m long tubes of the giant refractors had reached its peak. Gradually the balance shifted toward more stable and compact reflectors, instruments with only one optical surface rather than the four used with refractor objectives. Mirror-grinding techniques had also progressed; mirrors were now made of glass instead of metal and were coated with a highly reflective film of silver. Along with these developments, astronomers

■ *The Moon, photographed in 1900 by Maurice Loewy and Pierre-Henri Puiseux with the great .27 m Coudé equatorial telescope at the Paris Observatory.*

were also turning their attention beyond the planets and the stars, toward those mysterious "nebulae." More and more of these faint and fuzzy objects were being discovered, yet their size, nature and distance within the Universe were complete unknowns. Clearly much bigger telescopes were needed to study them.

THE AGE OF GIANT TELESCOPES

Modern cosmology developed in the 20th century. After the paradigms of quantum theory and general relativity were formulated, it had the capability to finally clarify the nature and evolution of the Universe. And yet, without the extraordinary advances in optical astronomy, the great cosmological questions would probably have remained purely academic. For example, in 1781 Messier's catalog of diffuse objects contained only 103 entries, but the *New General Catalogue of Nebulae and Clusters of Stars*, published in 1888 by Johan Dreyer

■ *Passionate about Mars, wealthy American amateur Percival Lowell (1855–1916) established an observatory in Arizona equipped with a .6 m refractor.*

The Hooker telescope at Mount Wilson Observatory in California was put into service in 1918. This instrument made history as the first telescope that permitted astronomers to observe at a true cosmological scale.

(1852–1926), listed some 7,840 objects. In the late 19th century, James Keeler (1857–1900) undertook a deep-sky photographic survey with the .91 m Crossley Reflector at Lick Observatory. He reached objects at 15th and 16th magnitudes, and estimated from this that more than 100,000 spiral nebulae were within reach of his telescope.

The big question, however, was still whether these wispy spiral nebulae were close to us or not. Were they relatively close, turbulent clouds of evolving planetary systems, or gigantic condensations of stars at enormous distances? Even more central, did the immense accumulations of stars within the Milky Way represent the entire Universe, as most astronomers at the time believed? Or, as philosophers Thomas Wright (1711–1786) and Immanuel Kant (1724–1804) had suggested, was the cosmos filled with other "island universes" similar to the Milky Way but very, very far way?

In the early 1900s, astronomer George Ellery Hale (1868–1938), father of the great 1 m refractor at Yerkes, set out to build a large modern reflector entirely dedicated to spectroscopy and photography. This resulted in the giant 1.52 m reflector of 7.6 m focal length atop Mount Wilson, California, in 1909. This telescope provided the first truly detailed photographs of spiral nebulae, using exposures up to 10 hours taken over four or five successive nights. For the first time as well, the 20th magnitude "wall" was breached. Hale continued along this fruitful path as director of Mount Wilson Observatory. Only six years later, in 1918, one of the most important astronomical instruments in history was put into service under his direction: the famed 2.5 m Hooker telescope with a 12.5 m focal length.

This modern telescope, on its yoke-type equatorial mount, registered objects a million times fainter than those visible by eye and provided the first glimpses of the Universe on a cosmological scale. Edwin Hubble (1889–1953) was able to resolve individual stars in some of the spiral nebulae with this superb instrument, and in 1924 he photographed giant "Cepheid variables" in the Andromeda Nebula (M31). This discovery was of tremendous importance. As Henrietta Leavitt (1868–1921) had demonstrated several years before, the maximum brightness of these variables is closely related to their periodicity. Consequently, since the absolute brightness of such stars was known, they could be used as "standard candles" by astronomers, allowing them at last to calculate the distances to spiral nebulae.

And so it was that philosopher Immanuel Kant was proven correct: the spiral nebulae, containing hundreds of billions of stars and several million light-years away, were indeed very distant island universes. Galaxies had been discovered.

Continuing to work with the Hooker telescope, Hubble soon discovered that galaxies are receding at velocities

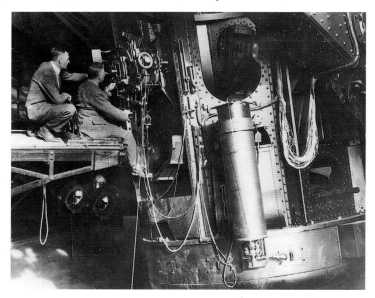

Edwin Hubble and Milton Humason at the focus of the 2.5 m telescope at Mount Wilson. This instrument enabled Hubble to discover the expansion of the Universe.

proportional to their distances. These observations were in line with the general theory of relativity and showed that the Universe is expanding (a fact Einstein himself ignored). This work eventually gave rise to the big bang theory of the origin of the Universe.

THE 5 M MOUNT PALOMAR TELESCOPE

In the chill of the evening of November 11, 1950, Edwin Hubble hurried through the thick pine forest surrounding the domes of Mount Palomar Observatory, located far south of Los Angeles at an elevation of 1,700 m. The technicians, returning home after a long day at work, moved past the tall figure of this most famous astronomer, who was draped in a long coat with a raised collar. Passing the smaller dome of the Schmidt telescope on the left, Hubble slowly climbed a series of paths leading to the largest telescope ever built. The thin crescent Moon, soon to set, dimly illuminated the great dome of the Hale telescope, rising more than 40 m above ground.

George Ellery Hale died a dozen

■ *Conceived in 1930 and placed in service in 1948, the celebrated Hale telescope is still in use today. The 5 m Mount Palomar reflector is among the most notable icons of 20th-century astronomy.*

years before his crowning achievement was finished. After the 2.5 m Mount Wilson reflector was pressed into service, Hale dedicated himself to the construction of the largest astronomical telescope in the world, his fourth and last. In 1928, he persuaded the Rockefeller Foundation to finance the construction of an ultramodern reflecting telescope with a 5 m diameter mirror, a focal length of 17 m and three different foci. Although delayed by World War II, this gigantic, 530-metric-ton equatorial telescope entered operation in 1948. Perched in the observer's cage more than 20 m above ground, Hubble gazed into the eyepiece at the soft twinkle of distant stars that no one before him had seen.

In 1950, astronomers were still using photographic plates. A 30-minute exposure on Kodak emulsion 103-aO was required to image a spiral galaxy like NGC 1073 and reach stars as faint as 23rd magnitude. Could Hubble have possibly foreseen that no equatorially mounted telescope larger than this giant reflector would ever be built? Or that by turn of the century the venerable Hale telescope, now only sixth largest in the world, would still be in active service?

Hubble could never have imagined that in the next 50 years, multimirror telescopes 8 to 10 m in aperture, controlled by computers and equipped with adaptive optics and electronic cameras, would provide images with six-fold higher resolution and a hundred times brighter than the one he was capturing at that moment.

Control paddle in hand and with his eye glued to the dimly lit reticle of his eyepiece, the famed American astronomer had to patiently track a guide star while carefully regulating the slow movement of his giant telescope. Could he possibly have imagined that in the future, high above the atmosphere, a space telescope of incredible capability would bear his name and detect billions of galaxies down to 30th magnitude, located near the confines of the big bang? ■

The Great Observatories

■ This impressive photograph shows but a small portion of the Milky Way in the constellation Sagittarius. The apparent density of stars in this image is somewhat misleading; in reality the average distance between stars is more than one light-year. More than 100,000 stars are visible in this image recorded by the 3.6 m Canada–France–Hawaii Telescope equipped with the CFH12K camera.

The Pic du Midi Observatory

FRANCE

As you travel south over the Lannemezan plateau or the plains of Bigorre toward the immense, snow-covered wall of the Pyrenees, one triangular-shaped peak dominates the mountain chain extending from Cerberus to Hendaye. This is deceiving, however. When approached from the north, Pic du Midi de Bigorre appears much larger than its neighbors because it stands some 20 km apart from the other major peaks, including mounts Néouvielle, Vignemale, Marboré and Perdu, all exceeding 3,000 m in elevation. Rising to 2,870 m, Pic du Midi itself is uniquely situated. Toward the south it borders a 200 km stretch of the Pyrenees. In the north it borders Gascony, draped in fog and summer heat. The Basque territories and Béarn lie west, and Lauragais in the east. At night Pic du Midi's dark sky rises far above the shimmering glow of distant lights that link distant towns and villages like a luminous spiderweb. These are the roads and highways of the towns of Bagnères-de-Bigorre, Lannemezan, Montréjeau, Tarbes, Pau, Toulouse, the lights of Lacq and, way in the distance, the lighthouse at Biarritz.

Pic du Midi observatory was built between 1878 and 1882 at the urging of a retired brigadier general, Charles de Nonsouty, and an engineer, Célestin Xavier Vaussenat. At the time, the mountaintop was accessible only on foot and mules carried all construction materials to the summit. The observatory began as an elongated, two-story stone structure, 8 m wide by 20 m long, protected by 1 m thick walls. By the beginning of the 20th century the exceptional astronomical qualities of the site were fully recognized and in 1904 Bejamin Baillaud, director of Toulouse Observatory, erected the first dome and a .23 m refracting telescope at the site. Astronomers soon found that image quality there was always good, often very good, and occasionally superb.

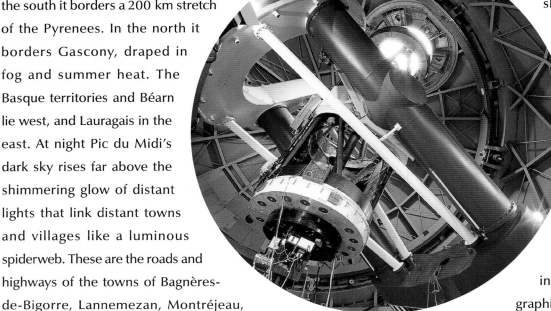

■ *The 2 m telescope at Pic du Midi was installed on a horseshoe-style, yoke equatorial mount. It is equipped with CCD cameras and a high-performance polarimeter to study stars in our own galaxy.*

This is due to several factors, including the complete separation of Pic du Midi from the rest of the Pyrenees, and the predominant westerly winds that drive air smoothly across the peak in laminar layers unperturbed by other mountains. This unique situation is akin to observing from space, often making for exceptionally stable images. At an elevation near 3,000 m, the air can also be exceptionally transparent, as inversion layers trap haze, dust, pollution and moisture lower down in the atmosphere. This often creates a "sea of clouds" below the observatory and extremely clear, calm skies above it. It also creates the impression that the facility is cut off from the world and floating in the clouds above the long wall of the Pyrenees.

Between 1930 and 1960 and before the beginning of the Space Age, Pic du Midi played a leading role in solar-system astronomy. French observers at the time studied solar-system bodies intensely, both visually and photographically. In 1930 Bernard Lyot, one of the 20th century's most productive astronomers, installed the first coronagraph on the peak, a special telescope he had just developed at the Paris Observatory. Utilizing an occulting disk to block the intense light of the solar disk, this instrument facilitated direct observation of our star's chromosphere and the spectacular prominences emanating from it. Prior to this, such features were visible only during a total eclipse of the Sun, which amounted to a few dozen seconds or at best a few minutes in any given year. The performance of Lyot's coronagraph was enhanced by the exceptional clarity of the sky at Pic du Midi. At that elevation the atmosphere is almost perfectly transparent, and the corona, which is nearly a million times dimmer than the solar disk itself, stood out clearly. The invention of the coronagraph also opened the whole field of solar

■ *The frequent exceptional atmospheric conditions at Pic du Midi Observatory are due both to its high elevation and its relative isolation from the rest of the Pyrenees Mountains. At 2,870 m, the air is dry, clear and often remarkably stable.*

physics and placed Pic du Midi at its forefront, a distinction this Pyrenees observatory holds to this day.

FROM THE SUN TO THE STARS

In addition to observing the Sun, astronomers at Pic du Midi also initiated detailed surveys of the rest of the Solar System. Since the .23 m refractor installed in 1908 was not up to the task, Bernard Lyot mounted a .38 m refractor with his corona-graph in 1941. However, like its predecessors, this telescope was not able to take full advantage of the superior seeing conditions at this location, and was subsequently replaced by a .6 m coudé-type refractor. Eventually a .5 m solar refractor was also added, which provided extremely high-resolution images of the Sun (down to 0.25 arcsec) and first revealed "granulation" in the Sun's photosphere, the result of rising gas convection cells. Finally in 1965, the celebrated 1 m telescope was installed specifically for planetary work. Over time, these instruments became legendary, thanks to the exception-ally steady seeing conditions at this site. Each instrument produced the finest images of solar-system objects in its time, revealing details on the Sun, Moon, Venus, Mars, Jupiter and Saturn not seen previously. The 1 m, for example, undertook

fundamental investigations of the atmospheres of Venus, Mars and Jupiter, and discovered several new satellites as well as the E component of Saturn's ring system.

Today, the role of Pic du Midi observatory has changed. Since far superior observing sites have been discovered at desert locations in the Andes and on Hawaii, and progress in optics and electronics has raised the performance of the largest telescopes to new levels, the instruments at Pic du Midi are no longer exceptional. Nonetheless, this facility remains true to its tradition. High-resolution solar and planetary observing continues, a coronagraph monitors our star continuously, the .5 m solar refractor continues research on solar granulation and the 1 m follows the meteorology of Mars, Jupiter and Saturn. In 1994, this historic telescope observed the collision of comet Shoemaker-Levy with Jupiter, as well as the passage of comets Hyakutake and Hale-Bopp across our skies in 1996 and 1997. The observatory placed its largest instrument into service in 1980: the 2 m Bernard-Lyot reflector. This telescope is equipped with a high-resolution spectrograph, a stellar polarimeter and an infrared camera, but now that the Solar System is increasingly explored by spacecraft, its mission has been redirected toward stellar and galactic research. ■

The Haute-Provence Observatory

Its dozen white domes border a forest of twisted oaks and overlook a vast limestone plateau that is often windswept by the mistral. At 650 m elevation, the climate is too harsh here for olive groves. Toward the north lies Mount Ventoux, snow-covered a good part of the year. Toward the south, beyond some scattered dwellings and the nearby town of Saint-Michel-l'Observatoire, lie the wooded, rolling hills of the Lubéron natural preserve. Beyond and hidden from view are the towns of Forcalquier, the stately Manosque and Oraison, all overlooking the Ganagobie plateau. Still farther the boulders of des Mées delineate the border of Durance. Seen from a catwalk atop the 2 m telescope's dome, the tallest structure around, the vast panorama is reminiscent of the birthplace of Jean Giono, whose name was given to an asteroid discovered here a few years ago. During the summer, cicadas and crickets accompany astronomers at dusk on their way to the telescopes after leaving the reception center, a large, cool country mansion.

The Haute-Provence Observatory (OHP) was founded in 1937 by the Front Populaire government. Construction was halted during World War II but resumed again after the liberation of France. The observatory's facilities grew progressively after that. The first telescope, 1.2 m in aperture, was put into service in 1943, the 2 m in 1958, and the 1.5 m in 1969. OHP rose to prominence in the decades after the war. As the premiere facility in France and most of Europe, it attracted many astronomers to the clear skies of Haute-Provence. In the following years, several additional instruments were installed at the 100-hectare site, including an .8 m telescope, a 1 m instrument belonging to the Observatory of Geneva and, finally, a .6 m Franco-Belgian photographic Schmidt.

■ The history of science will note that the first extrasolar planet was discovered around 51 Pegasi, at Haute-Provence Observatory in 1995. Between 1995 and 2002, some 20 additional exoplanets were discovered there with the 2 m telescope.

After this, however, OHP went through some uncertain and difficult times. Several international observatories with 4 m telescopes and larger were built around the world, at better southerly locations and at elevations of 2,000 to 4,200 m. In comparison to these larger international facilities at the cutting edge of astronomical research, Haute-Provence suddenly appeared quite modest. It would have been understandable had most OHP astronomers followed their colleagues to those larger facilities, since it clearly requires more than a 2 m telescope to study 25th-magnitude galaxies billions of light years away. However, despite working under less pristine skies and more turbulence than telescopes in Hawaii and the Andes, OHP very effectively adapted to its scientific and budgetary challenges by specializing in stellar spectroscopy.

Concerned solely with the analysis of light of all colors and wavelengths, spectroscopy is not as affected by adverse sky conditions as are other areas of research. Spectroscopy lies at the foundations of modern astrophysics and most other areas of astronomy. Without it, we would know nothing of the chemical composition of stars and nebulae, nor anything of galactic distances or the expansion of the Universe. Due to the pivotal nature of this field, OHP set out to become a world leader in the spectroscopy of stars and the galactic interstellar medium.

During the 1990s, in the face of even more intense competition from international observatories and the development of telescopes in the 6 to 10 m aperture range, OHP fell under threat of total closure. To avoid this the OHP administration decided to limit its efforts even more and focus on areas of research that the giant facilities could not engage in. The demand for observing time at the world's largest telescopes is so great that review committees literally parcel it out half a night at a time. In contrast to this, OHP decided to allocate several nights, or even a whole week, to teams of astronomers needing time to study long-term variables, for example, or for large-scale surveys of stellar populations in the Milky Way. This can be tedious work, often requiring tens or even hundreds of nights of observations, but it is crucial to fuller understanding of stellar evolution and galactic dynamics.

And then one fine autumn evening in 1995, a team of Swiss astronomers led by Michel Mayor announced the discovery of the first planet outside the Solar System. Using the Elodie spectrograph and the 2 m telescope, they had found a planet orbiting 51 Pegasi. This discovery rocked the astronomical community worldwide; many others had searched diligently for decades without success for such elusive "exoplanets."

THE VERY FIRST EXOPLANET

This spectacular discovery brought immediate and worldwide acclaim to Haute-Provence Observatory and assured its place in the history of astronomy. However this did not just happen by chance. The Elodie spectrograph, built by André Baranne at Marseille Observatory and installed on the 2 m telescope in 1995, was one of the most precise instruments of its kind in the world. It took such a long time to discover planets beyond our Solar System, something philosophers and astronomers alike have contemplated for two millennia, because even now this lies at the limits of our technology. Planets only radiate reflected light and are relatively small, making it extremely difficult to detect them in the glare of their parent star. A planet is typically a billion times dimmer than

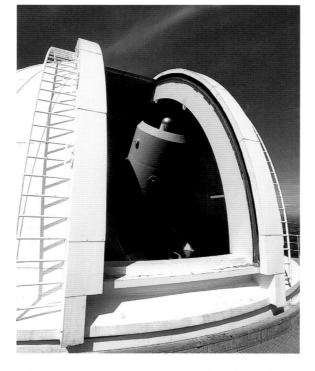

■ *The 1.5 m telescope is equipped with Aurélie, a spectrograph providing a resolution of 120,000. This instrument is used for observing stars and nebulae.*

the star it orbits. Since this makes it almost impossible at present to detect exoplanets directly, astronomers have resorted to indirect methods. The approach used by Michel Mayor and his team was to look for evidence of exoplanet-induced, gravitational perturbations in stars. The gravitational tug exerted by a planet causes slight, periodic wobbles in the star's motion across the sky, since the two bodies rotate around a common center of gravity. Changes of this nature can be detected by spectroscopy. In the case of 51 Pegasi, the invisible planet affected its radial velocity by 56 m/s, and translated into a Jupiter-size planet orbiting its parent star in only 4.23 days.

And so it was that in October 1995, with a 2 m telescope and its Elodie spectrograph, OHP won out over much larger instruments and began one of the most exciting scientific quests of our time. This was clear vindication for an observatory on the verge of closure, by showing that a modest, well-run national facility could be as effective as its distant cousins in Arizona, the Andes and Hawaii. The search for exoplanets has since intensified at OHP. Although a Franco-Swiss team continues working with Elodie, a new spectrograph, Sophie, will soon replace it. A third, high-performance spectrograph will also be installed on the 1.5 m telescope, increasing the possibility of additional exoplanet discoveries. To date some 20 exoplanets have been detected at Haute-Provence Observatory.

The planet orbiting 51 Pegasi has completed more than 1,000 revolutions since its discovery. OHP astronomers continue to monitor its progress with Elodie, accumulating more and more data that over time will help to reveal any additional planets. ■

■ *The spectrograph Sophie will replace Elodie on the 2 m telescope. Three times more accurate and 10 times more sensitive than Elodie, Sophie will lead to discovery of new exoplanets.*

The Calern Observatory

FRANCE

I t is almost a desert. Scattered yellow grass, whorls of thyme and twisted shrubbery struggle through arid summers and are often covered with snow by the middle of winter. Lost amid this subalpine terrain, the observatory's domes look tiny on the immense Calern plateau. At an elevation of 1,270 m, not far from the villages of Grasse and Saint-Vallier-de-Thiey, this remote site provides astronomers with more than 200 clear nights a year.

Astronomical telescopes were first installed here in 1974. At the time, the 20-square-kilometer site belonged to the Center for the Study and Research of Geodynamics and Astronomy (CERGA). In 1988 Nice Observatory joined it and the two were renamed the Observatory of the Côte d'Azur. This unusual facility serves as an optical laboratory and astrometric facility. Since it is involved in advanced research and development of new technologies, its astronomical mission is clearly long term.

A little over 20 years ago, a novel and very impressive project began at Calern Observatory. A 1.5 m telescope equipped with an atomic clock and a powerful laser was used for "lunar laser" measurements. These were coupled with another set of instruments quite far from Calern, on the surface of the Moon! A series of highly reflective panels, which were left on the lunar surface by the Apollo 14 and 15 missions and the Soviet Luna 21 probe, were used in combination with the laser-emitting telescope on Earth. The principle underlying these experiments is quite simple; pulses of laser light are reflected back from the lunar surface, detected by the telescope, and then the round trip time is measured with great precision by a cesium atomic clock. Since the speed of light in a vacuum is known to an accuracy of 299,792.458012 km/s, the distance between the Earth and the Moon can be determined to within a few centi-

■ The facilities on the Calern plateau are under the direction of the Observatory of the Côte d'Azur. At an elevation of 1,270 m, this site specializes in geodynamics and astrometric measurements. The dome of the 1.5 m lunar-laser telescope is shown here.

meters. This information is of considerable scientific interest since it can help detect any variations in the motion of the Earth and Moon. This, in turn, can provide information on gravitational perturbations between the two, as well as recording atmospheric friction and seismic events on Earth. Perhaps the most spectacular results of these lunar-laser ranging experiments are the confirmation that the Moon is gradually drifting away from the Earth at about 5 m per century and the validation of the theory of general relativity through some of the most accurate measurements ever made.

While simple in theory, obtaining precise measures of the Earth-Moon distance is complicated in practice. The laser emits a 300-picosecond pulse (0.3 billionth of a second!) every second, 10 times per second, which equates to less than .1 m in length. While each pulse is extremely powerful, it is also very brief. If it were a continuous emission, the laser would dissipate billions of watts of energy and destroy the telescope in the process! A major problem facing astronomers is that at the Moon's distance, the laser beam is spread over several square kilometers, which greatly diminishes its energy. Because of this, only one photon in every 100 trillion sent to the Moon is recovered 2.5 seconds later by the telescope's photomultiplier. A cesium atomic clock times the arrival of each photon to about 50 picoseconds. However, since photons travel an additional 15 mm in that 50 billionth of a second interval, the actual distance between the Earth and the Moon cannot be determined more accurately than that.

FROM THE EARTH TO THE MOON

The Cote d'Azur Observatory data are immediately converted to actual distance measurements by high-speed computers. This

is needed because even during the light's brief 2.5-second round trip, the positions of the Moon and Earth shift due to their respective orbital

■ *Calern's large, two-telescope interferometer is distinguished by its unusual architecture. The two 1.5 m telescopes can be moved back and forth along a baseline of 65 m. This instrument has made it possible to measure the diameter of several stars.*

motion and rotation. In fact, in a single night of observing, the position between the light's point of origin and its lunar target changes by several thousand kilometers. Space/time curvature effects due to changes in the Sun–Earth–Moon geometry must also be factored in, since these can deflect the laser beam by as much as 10 m.

A very large dome near the edge of the Calern plateau houses a totally different type of telescope used only on dark, moonless nights. This Schmidt telescope, a large photo camera with a 1.5 m mirror and a .9 m corrector plate, was installed in 1981 but is no longer in use today.

The observatory's most impressive instrument, however, is located between the domes of the Schmidt and the lunar-laser telescope. The silhouette of this facility appears like some strange, ancient monolith in evening twilight. Built during the 1980s, the GI2T (Giant Two Telescope Interferometer) does not resemble any other astronomical instrument and is considered a historic landmark even though it is still in active use today. The two 1.5 m telescopes at the core of the instrument are made of concrete. Designed by Finnish architect Antti Lovag, the two bowling-pin-shaped telescopes are mounted on rails that extend on both sides of this unusual laboratory, which was conceived by astronomer Antoine Labeyrie, a pioneer

in optical interferometry. Early in the 20th century, astronomers first contemplated combining light from two telescopes, and creating a virtual telescope with an effective aperture equal to the distance between them. Although the American astronomer Albert Michelson successfully measured the diameter of several bright stars during the 1920s using the interferometer at Mount Wilson Observatory, this was not subsequently continued because of the technical challenges involved. Antoine Labeyrie and his colleagues built a small-scale interferometer at Calern using .26 m aperture instruments prior to undertaking the full scale GI2T project in the 1970s. The two 1.5 m telescopes create an interferometer whose north/south baseline extends from 12 to 65 m, providing resolution from 0.01 to 0.002 arcsec in the visible range. This facility has carried out hundreds of measurements of stellar diameters, particularly red supergiants. GI2T is currently used to study the gaseous envelopes around variable stars and nearby interacting doubles like Beta Lyrae. It has been discovered that this close binary system is continuously exchanging matter, and that its 12.93-day rotation period is slowing about 20 seconds a year due to gravitational interactions. The pair will probably fuse in the future into the glow of a supernova. A third telescope will soon join GI2T, this time in an east-west direction along the plateau. The GI3T combination will let researchers construct full disk images of stars, complete with surface details. ■

> Light: the messenger of the stars

*A*stronomical objects have always intrigued us, even those we cannot see directly. Since most of them are forever beyond our reach, we must contend with deciphering the information inherent in the light (or more correctly, the electromagnetic radiation) that comes to us from the stars, planets, galaxies and interstellar domains. This has enabled us not only to measure the distances of these objects, but also their temperatures, composition and physical nature. The human eye is sensitive only to the narrow, visible portion of this extraordinarily rich electromagnetic spectrum, but astronomers continue to develop improved methods of capturing and studying all its wavelengths.

THE NATURE OF LIGHT

Most features and properties of light are well explained by the "wave" theory. This notion, first introduced by Huygens at the end of the 17th century, describes light as electromagnetic waves propagated in accordance with the laws formulated by Maxwell in 1872.

The wave theory, however, did not adequately explain all properties of light. One example is the photoelectric effect, whereby certain substances emit electrons when illuminated or exposed to electromagnetic radiation. To explain this effect, Einstein suggested in 1905 that light consists of discrete energy particles, or photons, similar to the "particulate" theory of light, first proposed by Newton in the 17th century but then abandoned in favor of the wave theory.

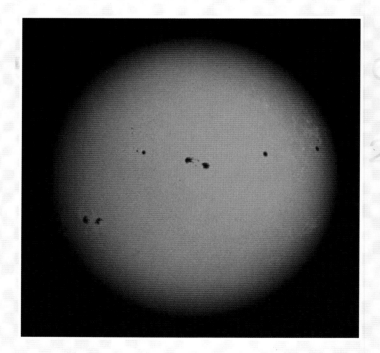 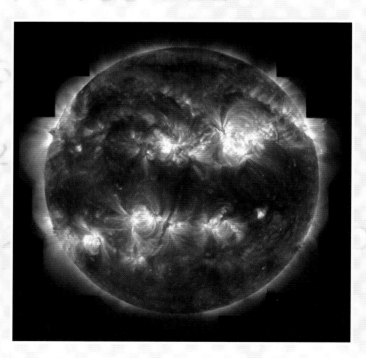

● Although celestial objects emit energy across the entire electromagnetic spectrum, most radiate strongest at wavelengths indicative of their inner workings. The Sun's maximum energy output, for example, is in the visible range of the spectrum (at wavelengths around 550 nm, image on left), which also happens to correspond to the maximum sensitivity of the human eye. An X-ray image of the Sun is shown at right, taken at wavelengths between 17 and 28 nm by the TRACE satellite.

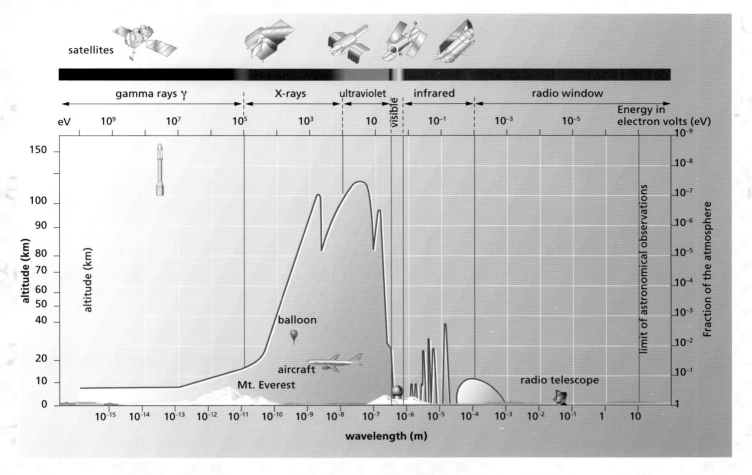

● The electromagnetic spectrum extends from very high frequencies (or very short wavelengths) to very low frequencies (or very long wavelengths). These broad ranges are further subdivided into smaller domains: for example, soft and hard X-rays; extreme ultraviolet, far and near; and for radio, submillimeter, millimeter, centimeter, decimeter and meter, wavelengths, or very high frequencies (VHF), medium (MF) and low frequencies (LF).

Are these two concepts of the nature of light, waves and particles, incompatible? In 1920, Louis de Broglie showed that they are not, and that light exhibits properties he called the "wave-particle duality." The dual nature of light was subsequently explained in terms of quantum mechanics.

THE ELECTROMAGNETIC SPECTRUM

Photons can be pictured as carriers of electromagnetic energy, each discretely defined by energy levels corresponding to the wavelength (or equivalent frequency) of the radiation in question. The electromagnetic spectrum comprises the full range of energy domains or wavelengths, and extends from gamma rays (extremely high energy or very short wavelength) to radio waves (low energy, long wavelengths), with X-rays, ultraviolet, visible and infrared wavelengths in between. In addition to intrinsic energy differences, spectral domains are also defined in practice by the methods used to detect them, although there are no fundamental differences in the nature of electromagnetic radiation, which is uniform across the entire spectrum.

Astronomers have probably focused so much attention on the visible range of the spectrum (between 0.4 and 0.7 μm) because that's what the human eye is most sensitive

to. Other spectral domains, however, are just as complex and informative. For example, highly energetic objects and processes, such as black holes and close, interacting binaries, emit X-rays and gamma rays. The composition of stellar atmospheres and molecular clouds are best studied in the infrared and millimeter ranges, while information on their atomic and ionic makeup can be obtained in visible, infrared and ultraviolet light. Measurements at several different wavelengths are usually needed to accurately assess the masses and temperatures of stars and other structures. In fact, modern astrophysics depends on observational data obtained by instruments working in various regions of the spectrum. We can draw an analogy here with information obtained on the human body. In visible light only the exterior details are revealed. Infrared images highlight some underlying layers, but X-rays are needed to reveal such interior structures as bones and organs. For a complete picture, all such information must be combined.

Access to the traditionally "invisible" portions of the spectrum is a relatively recent development in astronomy, and has profoundly influenced our overall view of the Universe: its structure, composition and the large-scale physical phenomena within it. For example: the cosmic background

● The appearance of the spiral galaxy M81 differs drama-tically depending on the wavelength used to image it. From top to bottom, ultraviolet light (between 200 and 280 nm), visible light (between 350 and 600 nm) and in the near infrared (between 1.25 and 2.16 μm). Hot, young stars radiate strongest in the ultraviolet, while older and cooler stars are prominent in infrared.

radiation detected at radio wavelengths provided direct support for the big bang theory; infrared signals have helped clarify our understanding of the birth and death of stars; and, most recently, the Chandra orbital X-ray observatory provided evidence for a supermassive black hole at the core of the Andromeda Galaxy.

FROM LIGHT TO ASTRONOMICAL INFORMATION

Electromagnetic radiation from astronomical objects generally travels very long distances before a telescope collects and analyzes it. In the absence of any intervening matter, light travels through space in a straight line at a velocity approaching 300,000 km/s. However, any interfering medium, including our atmosphere, interstellar dust and other astronomical bodies, can attenuate its intensity and affect its path. To be effective, the collecting areas of optical and other telescopes must be large enough to match the wavelength of the radiation under investigation. For a typical radio telescope, this translates into several hundred square meters, for an optical telescope some tens of square meters, and for an X-ray telescope, less than one square meter.

Once a telescope captures light, it usually directs it and focuses it on whichever suite of optical and mechanical instruments researchers need to get the information they want. Typically this includes cameras equipped with filters for imaging at various wavelengths and spectroscopes to disperse light into its spectral components via prisms or gratings. With interferometry, where several telescopes may be used to observe the same object, radiation from individual instruments must be combined to obtain an image or other usable information.

Photometers are another type of analytical instrument used by astronomers. Through a variety of photosensitive materials, photometers convert light energy into measurable electric currents or resistance. Depending on the type of photoreceptor used, the energy is converted into thermal, photochemical, photoelectric and other reactions, which can be measured and quantified. Once again, the type of receptor required is dictated by the wavelength of the radiation analyzed.

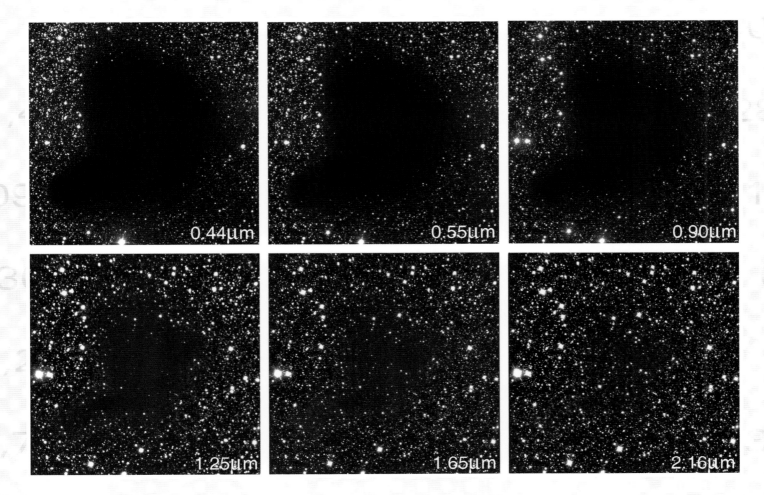

● Illustrations showing the dark nebula, Barnard 68, located 500 light-years from Earth. It is imaged at different wavelengths by the 8.2 m Antu telescope (top row) and the 3.5 m NTT (bottom row). In visible light at 0.44 μm, absorption by gas and dust reaches 35 magnitude and the cloud is totally opaque. This changes with increasingly longer wavelengths, and in the infrared, the cloud is totally transparent.

The first receptor used in optical astronomy, apart from the human eye, was the photographic plate. This photochemical technology capitalized on the ability of light to trigger chemical reactions in photosensitive materials and generate an image. Today CCDs (charge-coupled devices) have largely replaced the photographic plate. They use photoelectric materials that release electrons when interacting with light. In the high-energy domain of gamma and X-rays, scintillation detectors and counters are required, and other types of detectors are used for infrared, submillimeter, gamma and X-ray astronomy. These include bolometers, which measure light-induced temperature changes in gases and crystals. The signals and data acquired by various astronomical instruments are typically stored on magnetic tape or optical disks. At this stage the raw data contain useful information as well as artifacts and noise introduced by the atmosphere and detector defects, or in the case of solar astronomy, by the telescope itself.

Such factors can limit the usefulness of the observations and occasionally invalidate them completely. The final stage in the process is the reduction of data to extract the most meaningful and reliable results. In most cases, however, data reduction cannot remove all noise from the results and a measure of uncertainty or error remains. Whether done automatically by computers or manually by the observer, data analysis is usually carried out later and away from the observatory.

The long process of data analysis is particularly important in areas of astronomy where precise quantification or measurements are needed. This may require not only qualified astronomers, but also professionals in such fields as solid-state physics, optical engineering, electronics and mechanics. In addition, highly qualified technicians are needed to operate the telescopes and any auxiliary instruments necessary for data reduction and analysis. That is why it often takes many years between conceiving and designing a new instrument and its actual deployment.

Other important but more specialized fields of modern astronomical research include direct investigation of the Solar System itself, neutrino astronomy (particles produced through interactions of both strong and weak forces) and the search for gravitational waves (waves generated by massive, accelerating bodies as predicted by the theory of general relativity). ∎

The Plateau de Bure Observatory

FRANCE

The rounded backs of the six large antennas serve as breaks against winds that often exceed 100 km/h. The antennas are locked down in a horizontal position, their smooth carbon-fiber surfaces facing east and their backs turned into the blowing snowstorm until it passes. Located in the heart of the Alps between Grenoble and Gap, this expansive glacial plain rises to an elevation of 2,550 m and is enclosed by a wall of high mountains. This site offers astronomers the most pristine skies in Europe. The atmosphere is extremely dry and ideally suited for the array of instruments that occupy a significant portion of the plateau.

As soon as the storm has passed, the Bure interferometer faces skyward again and points at a young planetary system in the constellation Orion, or perhaps toward Taurus to monitor the turbulent early stages of star formation, or in the direction of Virgo, at a galaxy near the edge of the Universe.

Radio telescopes working in the millimeter range can observe some of the coldest regions in the Universe. They can penetrate areas that are either opaque in visible light or too cold to emit any visible light. The Milky Way's

■ *The Plateau de Bure interferometer specializes in origins, focusing on star birth in the Milky Way and the formation of galaxies in the early Universe.*

dust-rich molecular clouds, for instance, appear opaque in visible light, yet are totally transparent at millimeter wavelengths.

The Institute for Millimeter Radio Astronomy (IRAM) installed the first three antennas of the Bure interferometer between 1986 and 1988. At that time, the millimeter range of the electromagnetic spectrum, between the infrared and the longer wavelengths of traditional radio astronomy, was not well characterized. Research in this area was just beginning, and only a few large facilities existed working at this wavelength.

France, Germany and Spain established IRAM in 1979. A staff of about a hundred, based at the University of Saint Martin-d'Hères near Grenoble, oversaw two observatories, one at Bure and the other on Pico Veleta in Andalusia. The latter, located high in the

Sierra Nevada, began using a 30 m dish in 1985 to study interstellar clouds in the Milky Way.

The Milky Way's dark nebulae are relatively dense and extremely cold. They contain between 1,000 and 10,000 atoms of matter per cubic centimeter, at temperatures around 10 K (–263°C). Deep within these clouds, complex interstellar molecules can form and grow, shielded from the intense ionizing radiation of nearby young, supergiant stars, whose destructive energy would quickly dissociate them again into simpler atoms. These dark clouds are like chemical laboratories to astronomers. Over 100 different types of molecules have been identified in them, including some 20 species not present on Earth. Fortunately, carbon monoxide (CO), a strong emitter at millimeter wavelengths, is also common in these regions, making it an ideal "tracer" molecule for hydrogen, a much less reactive gas. Since there are about 100,000 hydrogen molecules (H_2) for every CO molecule in the Milky Way, tracking carbon monoxide is a good way of measuring both the amount and distribution of hydrogen, the most abundant element in the Universe.

With a collecting area of more than 700 square meters, the 30 m Pico Veleta antenna can detect very faint signals. Its resolving power, however, is rather limited. The resolving power of any astronomical telescope (and hence the clarity of images obtained) is a function of both its aperture and the wavelength at which it operates. This translates into less than 30 arcsec resolution for a 30 m dish working in the 3 mm wavelength range. To resolve detail comparable to that of modern optical telescopes, say 0.5 to 1 arcsec, this type of radio telescope would have to be several hundred meters in diameter.

Fortunately, it is possible to overcome this inherent limitation of radio telescopes through interferometry. This is achieved by combining signals from an array of several small antennas to

■ *The six dishes of the IRAM interferometer are located on the immense Bure plateau in the Alps at an elevation of 2,550 m. The array can be configured in several ways. The largest is 232 x 408 m and provides the highest resolution.*

simulate a receiver with a much larger effective aperture. That's the underlying principle of the Bure facility. Each dish in the array acts like a segment of a much larger radio telescope. Computers combine and integrate simultaneous signals from each dish to obtain higher resolution images. By adapting and refining this novel technique to shorter wavelengths, IRAM astronomers were able to improve the quality of their data in spectacular fashion. The interferometer, which began with only three antennas in 1990, now consists of six dishes with a collecting surface exceeding 1,000 square meters and has exceptional sensitivity. In fact, the resolution of the millimeter wavelength images obtained with this instrument is around 0.5 arcsec, comparable to that obtained by the best optical and infrared telescopes. This is equivalent to resolving detail on the Moon about 900 m in diameter, or 40 light-minutes at a distance of 30 light-years.

THE FORMATION OF PLANETARY SYSTEMS

In the 1990s the Bure interferometer undertook a systematic study of proplyds (proto-planetary disk sources), the small, opaque globules first observed in the Orion Nebula by the Hubble Space Telescope. Since these structures, which are about the size of the Solar System, are transparent at millimeter wavelengths, the Bure interferometer characterized them more fully. IRAM established that the largest of these proplyds contained about 1 percent more dust than the total mass of the Sun. This was an important discovery, since it is estimated that at least that much dust is required to form planets around a star.

Because the IRAM facility is one of the most powerful millimeter telescopes in the world, researchers there have extended their focus beyond embryonic star systems in our galaxy and are investigating some of the most distant galaxies in the Universe.

A review panel at IRAM invites research proposals from astronomers each year and then allocates observing time on the interferometer to the proposal deemed most meritorious. This works out to about 100 projects annually. The French, German and Spanish astronomers who are granted observing time at the facility will not actually go to Bure personally, as they might do with an ordinary telescope. That is because several observing runs may be needed over weeks or even months in order to obtain the best possible images of the objects under investigation. In addition, the configuration of the antennas is changed regularly by moving them along a pair of crossed rails 232 x 408 m long. This provides a larger area and a less expensive way of configuring

The six 15 m diameter dishes have a combined collecting area that exceeds 1,000 square meters. This produces a very large simulated radio telescope with spatial resolution of 0.5 arcsec, and equals the performance of the largest optical telescopes today.

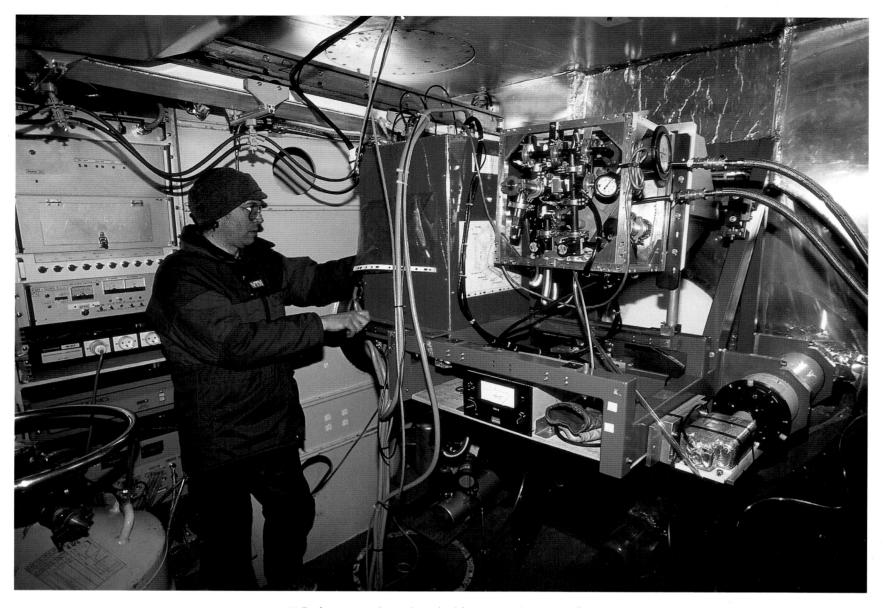

■ Each antenna is equipped with two receivers, one for 1.3 mm and one for 3 mm wavelength radiation. Both are continuously cooled to 2.8 K (–270.3°C). The next generation of receivers will permit observations down to wavelengths of 0.8 mm.

the interferometer array. A typical observing run at Bure involves at least three sessions lasting eight hours each. The telescope follows its target as soon as it has risen and then follows it until it sets. This continues around the clock, since observations at millimeter wavelengths are not affected by night or daylight. Approximately every 12 hours the five instrument operators meet in the control room and pool and process their data to generate a high-resolution image.

A BILLION YEARS AFTER THE BIG BANG

While molecular clouds and the birth of stars in our galaxy are the major focus of the Bure interferometer, it is also being used to study the far reaches of the Universe. Millimeter astronomy is probably the most powerful tool currently available to astronomers interested in the earliest, most distant structures in the Universe.

Researchers at Bure Observatory have detected both interstellar dust and the signature molecule CO in such galaxies as F 102114+4724, H 1413+117, and BR 1202-0725. These objects have redshift values ranging from $z = 2.3$ to 4.7, indicating that they formed when the Universe was between 10 and 15 percent its present age. If the Universe is about 15 billion years old, then quasar BR 1202-0725 in Virgo dates back to about one billion years after the big bang. The detection of interstellar dust in this quasar took the international astronomical community by surprise. This indicates that several generations of stars had already existed by that time and converted their primordial hydrogen and helium to heavier elements like oxygen and carbon. The latter were subsequently expelled into space through supernova explosions or by red giants shedding their shells. This implied that the earliest stars in the Universe must have formed well before then, probably only 50 million to 100 million years after the big bang.

It is likely that interferometers working in the millimeter range, rather than optical and infrared telescopes, will provide information on the furthest reaches of the cosmos. That's because the light generated by the bright young stars in the earliest galaxies has been absorbed by intervening gas and dust, and hence is undetectable by conventional optical telescopes. At the time they existed, these objects radiated mainly at infrared wavelengths between 60 and 100 μm. With the expansion of the Universe, however, this radiation has been shifted well into the millimeter range, and so is readily detected by the interferometer at Bure Observatory. ■

The Calar Alto Observatory

SPAIN

Viewed from the city of Almeria on the Andalusian coast, the Los Filabres range to the north looks like a huge, craggy wall dominated by the peak of Calar Alto. Even after most of the mountain snows have melted in the spring, several tiny white pearls remain dotting the summit: these are the domes of an observatory operated jointly by Spain and Germany. The Andalusian Mountains are an exceptional location for astronomy. The region, an arid semidesert, is located at the same latitude as California, New Mexico and Texas, with matching pristine skies. At an elevation of 2,160 m, Calar Alto provides astronomers about 200 clear nights a year. The observatory was established under the auspices of the Max Planck Institute for Astronomy (MPIA), which in the 1960s was looking for a northern research facility to complement the European Southern Observatory (ESO), in which Germany and France had a majority stake. German astronomy west of the Berlin wall was effectively nonexistent at the time. After all, the country's largest telescopes, the 1 m at Hamburg Observatory and the .72 m at Heidelberg, while still operational, were built in 1906 and 1910 respectively! Things were no better in East Germany. Although the world's largest Schmidt telescope, with a 2 m mirror, was completed in 1960, this Zeiss-built instrument was located on a forested hill near the city of Jena. Because of this, and the fact that the number of clear nights per year in Northern Europe suitable for astronomy rarely exceed 50, this facility was never able to perform at peak capacity.

■ *The domes of Calar Alto Observatory, illuminated by the setting Sun, dominate the summit of the Sierra de Los Filabres. Calar Alto provides German and Spanish astronomers with about 200 clear nights a year.*

THE LAST GREAT EQUATORIAL TELESCOPE

Construction of the observatory at Calar Alto began in 1973 and was completed about a dozen years later. Remarkably free of light pollution, the site covers an area of about 100 square kilometers. The first instrument to enter service there in 1975 was a modest 1.2 m equatorially mounted telescope weighing about 15 metric tons. A 1.5 m and a 2.2 m telescope were installed in 1979, and a .8 m Schmidt saw first light in 1981. In 1985, the last phase of work was completed with installation of the 3.5 m telescope. The 1.5 m is the only Spanish instrument at the site. Built by the firm of Reosc, it is under the direction of the National Observatory of Madrid. The .8 m Schmidt was built by the Carl Zeiss company and initially used at Hamburg Observatory before relocation to the superior skies at Calar Alto. The remaining three German telescopes, all built by the prestigious Zeiss firm, are similar in design. All are in the Ritchey-Chretien configuration, which provides wide fields of view with few optical aberrations. The equatorially mounted 2.2 m telescope weighs a massive 72 metric tons. It is one of a pair of telescopes, with the other located in the southern hemisphere at La Silla Observatory. Only the Calar Alto telescope, however, is equipped with a high-performance spectrograph with a resolution of 100,000.

The largest telescope at Calar Alto holds a special place in the annals of optical astronomy. This elegant, 230-metric-ton instrument is the last major one built under the old school of design. With its classic monolithic mirror of f/3.5 focal ratio and a horseshoe-style yoke mount, the design of this telescope mirrors that of the venerable 5 m Palomar giant conceived in the 1930s.

Ground and polished by Zeiss, the 3.5 m mirror is made of Zerodur, a ceramic material with an exceptionally low coefficient of expansion of 10^{-7}. The mirror exhibits great thermal stability under a variety of conditions, expanding or contracting less than one ten-thousandth of a millimeter for every 1°C change in temperature. This telescope is the flagship German instrument in the northern hemisphere. It is equipped with several high-performance accessories, including the Mosca spectrograph, and a widefield Laica camera with four CCD elements of 4,000 x 4,000 pixels each, covering a 45 arcmin field of view.

■ *The famed optical firm Zeiss built the 3.5 m reflector at Calar Alto. The last of the great equatorially mounted instruments, it was modeled after the venerable Hale telescope at Mount Palomar.*

The observatory at Calar Alto has excelled in a number of areas of research. German astronomers have specialized in the study of stars in our own galaxy, leading to the discovery of brown dwarfs at great distances from us, including star clusters in Taurus and Gemini. Other work has focused on the age of star clusters, and long-term studies involving several telescopes have been undertaken on the chemical composition of the interstellar medium. In the area of cosmology, project CADIS (Calar Alto Deep Imaging Survey) is examining remote galaxies in a sector of sky at high galactic latitude, which avoids interference from stars and nebulosity in the Milky Way. This survey has recorded objects fainter than 23rd magnitude at 40 different wavelengths, and has required hundreds of hours of observing time with both the 2.2 and 3.5 m telescopes. Its main goal is to provide MPIA astronomers with a contiguous "sample" of the Universe, reaching back billions of years and providing an accurate picture of the evolution of galaxies over time. By 2005 or 2006, this survey will probably be extended as far back as the big bang itself, once the much bigger, Large Binocular Telescope on Mount Graham in Arizona is brought on line. This huge telescope is located at roughly the same latitude as the Andalusian facilities, and was also financed in part by the Max Planck Institute of Astronomy. ■

The Pico de Teide Observatory

SPAIN

Rising 3,718 m above sea level, the striking pyramid-shaped volcano called Pico de Teide dominates the huge Canary Island archipelago. Like their faraway Hawaiian counterparts, the seven Canary Islands are of volcanic origin and located south of the Tropic of Cancer. Also like Hawaii, two of the Canaries provide ideal locations for astronomy. Extending 2,426 and 3,718 m above sea level respectively, the summits of La Palma and Tenerife rise above the cloud layer. Crossing the Atlantic Ocean from the west, clouds typically lie alongside the volcanoes' flanks below the 2,000 m level, creating ideal atmospheric conditions for astronomy. La Palma Observatory is dedicated to the night sky, while Pico de Teide has become a virtual "temple" of solar astronomy.

Scottish astronomer Charles Piazzi Smith was the first to discover the superb skies of Tenerife. In 1856, he persuaded the British Admiralty to send an expedition to this volcanic island and demonstrated that astronomical observation was much improved when positioned above the lowest layers of the atmosphere. With approval from Spanish authorities, Smith installed a small telescope on the island in the summer of 1856, first at 2,717 m atop Guajara, and later at 3,260 m on the southern flank of Pico de Teide. A number of expeditions to Tenerife followed in subsequent years, including one by Jean Mascart of the Paris Observatory to view Halley's comet in 1910. Won over by the exceptional skies of Tenerife, he was first to suggest that an international observatory be established on the Canaries. In the following years Mascart worked the Spanish, French and German diplomatic circles in this effort. However, his initiatives fell by the way in 1914 with the onset of World War I. Spanish astronomers revived this effort for the total eclipse of the Sun in 1959 and estab-

lished an observatory on Pico de Teide in that year. In the 1970s, solar photographs were obtained at Teide equal in quality and resolution to those of Pic du Midi, and this prompted renewed discussion about establishing a major observatory on the island.

Astronomers visiting Pico de Teide for the first time are invariably struck by the many contrasts they encounter on Tenerife: the airport of Santa Cruz sitting at the edge of a very jagged coastline, throngs of tourists from all over Europe seeking its perpetual spring-like climate, and en route to the summit, the flanks of a volcano covered by a dense forest of pine, cedar and oak. Finally, at the 2,400 m level, they reach the huge Teide caldera, a desert landscape of ocher and black lava. There, amid the rarefied air of Izana hill, they see the observatory domes.

In many ways solar observing is the antithesis of nighttime astronomy. At night astronomers struggle to collect, amplify and analyze the faint signals of dim, distant objects. In contrast, solar astronomy usually contends with too much light. Nighttime astronomers are always wanting bigger-aperture telescopes, while solar astronomers must filter out most of the Sun's light and intense heat. While night-sky telescopes must be as open as possible to reach thermal equilibrium with the night air, solar telescopes must be fully insulated against heat and solar radiation.

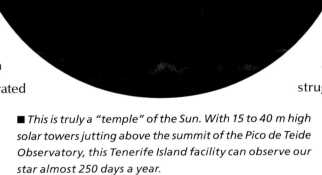

■ *This is truly a "temple" of the Sun. With 15 to 40 m high solar towers jutting above the summit of the Pico de Teide Observatory, this Tenerife Island facility can observe our star almost 250 days a year.*

THE TEMPLE OF THE SUN

The five large solar telescopes at Pico de Teide are .25, .4, .45, .6 and .9 m in diameter, and were built over several years. Each instrument has been optimized to reveal the finest solar detail possible. The best solar images obtained (often with adaptive optics) at Pic du Midi, Sacramento Peak, La Palma and Tenerife

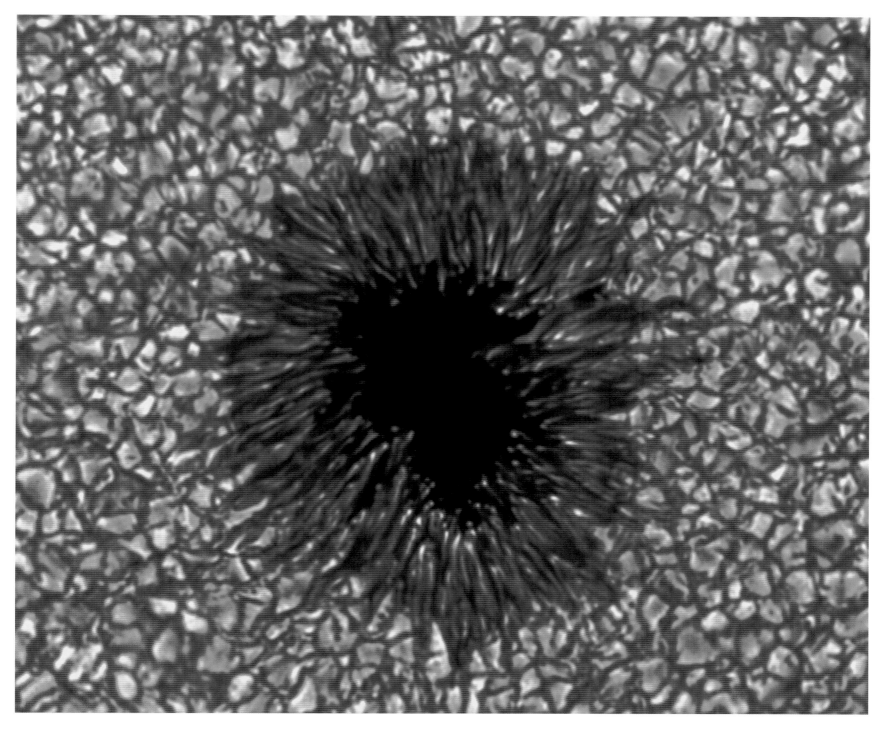

■ This very high-resolution image of the Sun's surface shows a sunspot close up. Sunspots appear darker because they are slightly cooler than the surrounding regions. Astronomers at the Astrophysical Institute of the Canaries took this photograph with the .5 m Swedish SVST, attaining .25 arcsec resolution.

have reached between 0.15 and 0.3 arcsec resolution, about the best possible with .4 to .9 m telescopes. This is also the best that can be expected from ground-based instruments. At that level of resolution, detail of only a few hundred kilometers can be detected on the Sun. To minimize atmospheric turbulence, all instruments at the observatory are housed in domed towers extending 15 to 40 m above ground. In addition, to minimize the extreme effects of solar heating, all telescopes, spectrographs and recording instruments are kept sealed under vacuum.

Sponsored by the Kiepenheuer Institute for Solar Physics, German astronomers installed three solar telescopes at Tenerife in 1989, including the .4 m Vacuum Newton Telescope, a .45 m Gregory-Coudé and the .6 m Vacuum Tower Telescope. However, French and Italian astronomers brought the largest and most innovative telescope to the site: Thémis. Installed on a 30 m tower in 2000, this .9 m vacuum telescope took several years to develop. With its 0.2 arcsec resolution capability, Thémis will address several key questions in solar physics, all relating to how powerful magnetic forces in the Sun affect the dynamics of superheated gases at various levels in the solar atmosphere, the photosphere, the chromosphere and the inner corona.

Attracted by the superior seeing conditions at Pico de Teide, nocturnal astronomers have also brought a number of telescopes to this location. As a result, several conventional domes have now appeared near the five towers of the "temple of the Sun," including .5 and .8 m telescopes, as well larger 1 and 1.5 m instruments. In addition, Italian, British and Spanish researchers have installed widefield radio detectors at Tenerife for work in the .01 to .03 m wavelength range. These will be used for research on the cosmic background radiation emitted about 300,000 years after the big bang. ■

> Telescopes, mountings and mirrors

The telescope is only the first in a series of elements needed to collect and analyze radiant energy from celestial objects. Because of this, constant efforts are made to improve and enhance telescope performance. Modern telescopes are true technological marvels, something unimaginable just a few decades ago.

COLLECTING LIGHT

The primary function of telescopes is to collect and direct light rays toward support instruments that will filter or disperse them before reaching the final detector. In the optical and infrared range, photons are usually collected by a primary mirror, which is either parabolic (Cassegrain design), hyperbolic (Ritchey-Chrétien) or spherical (Schmidt telescope). This mirror reflects and brings light to either a "primary focus" or, more commonly,

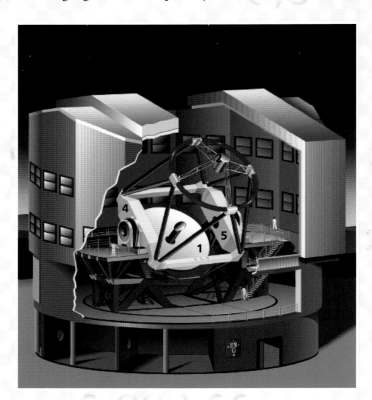

• A telescope's primary mirror (1) captures the light of astronomical objects. The light is concentrated toward the secondary mirror (2), which redirects it toward one of several foci. A modern telescope like the European VLT typically has three foci: a Cassegrain (3), located behind the perforated primary mirror, and two Nasmyth foci (4 and 5).

toward a smaller secondary mirror. The latter in turn sends the beam toward a "secondary focus," where another flat mirror sends it toward either a Nasmyth or Coudé focus.

The type of mount selected for a given telescope depends on such factors as its optical configuration, the number and types of foci it has for specific astrophysical applications, and of course, cost. For example, despite some initial difficulties, widefield imaging is usually done at the primary focus. The spatially fixed Coudé focus is well suited for instruments requiring large image scale or high sensitivity, such as large spectrographs.

Multimirror telescopes combine several smaller mirrors configured to work as one. With their unusual focal configuration, Cassegrain telescopes require a special type of mounting, but offer the advantages of a really large aperture telescope at a relatively low cost. They are, however, somewhat difficult to bring into full operation.

TELESCOPE CHARACTERISTICS

Whatever their configuration, all telescopes share some basic characteristics and parameters. The first of these is diameter or "aperture," which dictates how much light or radiation a given instrument can capture. This is a function of the surface area of the mirror or collector, or approximately the square of its diameter. The diameter of a mirror also determines its resolution limit, or its ability to separate objects that appear close together. For any given wavelength, the resolution limit is inversely proportional to the telescope's aperture. Another key characteristic is the instrument's focal ratio, sometimes also called "aperture ratio," which is equal to focal length (f) of the optical system, divided by the diameter (D) of the primary mirror. The shorter the f-ratio (f/D), the "faster" the

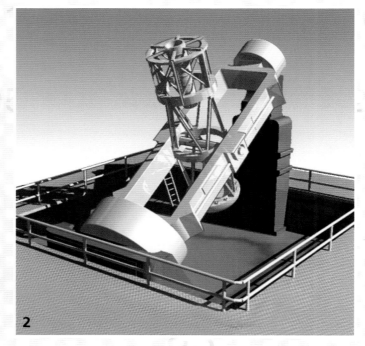

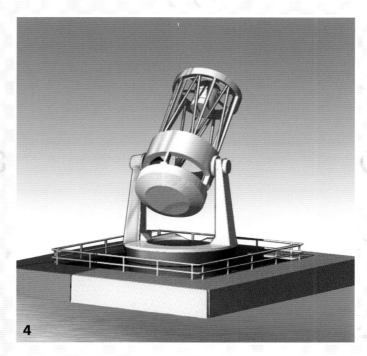

● Even toward the end of the 20th century, most telescopes were still installed on equatorial mounts. Despite their differences in style, the English (1), yoke (2) and horseshoe type (3), were based on the same principle; they followed the apparent motion of celestial objects by rotating around a polar axis parallel to the Earth's axis. Tracking is at the exact same speed, but in a direction opposite to the rotation of our planet. Today telescopes are installed on much more stable altazimuth mounts (4). Computers must control three directions simultaneously with altazimuth-mounted telescopes: the vertical and horizontal axes, and the rotation of the instrument itself at the focal plane.

system is said to be in terms of converging the cone of light rays. For any given aperture, the smaller its f-ratio, the shorter a telescope will be. Even if the f-ratio of the primary is short, it will be larger at the Cassegrain or Nasmyth foci, and very large (f/40 to f/50) at the Coudé focus. The sky coverage or field of view of an optical system is typically expressed in arc seconds or minutes. The corresponding area on the surface of a CCD or other receptor is determined by the scale of the image at the telescope's focal plane, and usually expressed in millimeters. These dimensions also depend on the telescope's focal length.

When construction of a major new telescope is planned, all of the above factors are taken into consideration. Once such parameters as aperture, focal ratios, field of view and image quality are specified, the most appropriate optical configuration is selected.

TRACKING CELESTIAL OBJECTS

As is well known to experienced observers, celestial objects appear to move across the sky due to the Earth's rotation about its axis. To compensate for this, telescope mounts are equipped with electric drives that track their target during extended periods of observation. Unless such tracking is very precise, the object will gradually move out of the field of view during long photographic exposures. Telescopes must be mounted not only for stability and support, but also to point freely in any direction in the sky. Consequently, both the equatorial and altazimuth types of mounts feature two axes of rotation perpendicular to each another.

GETTING THE MOST OUT OF IMAGES

In addition to capturing photons, the most important requirement of any telescope is the ability to produce quality images. Many different factors affect this. Atmospheric conditions play a major role, at least for ground-based telescopes. Other key elements are the types and quality of optics involved and the instrument's tracking accuracy and stability.

To ensure good image quality, the optical elements in a telescope must be perfectly aligned and collimated, and free of chromatic aberration, coma, spherical aberration, astigma-

● This picture shows the mirror support of the 3.5 m WIYN telescope at Kitt Peak Observatory. The back of the mirror is equipped with numerous actuators whose orientation changes depending on the direction the telescope is pointed. These computer-controlled devices (active optics) can regulate and deform the shape of the mirror and significantly improve image quality.

tism and similar optical defects. Ray tracing is an important application in telescope design. This technique delineates light

● The mirror of the Gemini-North telescope is 8.1 m in diameter, 2 cm thick and weighs 22 metric tons. Its parabolic figure has been corrected to an accuracy of one-fifteenth nm. It is unlikely at present that a much larger, solid mirror can ever be cast, ground and polished. To get even larger aperture telescopes, like the Large Binocular Telescope on Mount Graham, astronomers must combine several mirrors, in this case two 8.4 m mirrors. Another alternative is to build segmented mirrors, like the two 10 m Keck telescopes, each consisting of a mosaic of thirty-six 1.8 m mirrors.

rays in the optical path and shows where they are brought to focus. Typically, the smaller the blur point at the focal plane, the better the optics. Ray tracings are also used to ensure good image quality across the telescope's entire field of view.

It is very important that all optical surfaces in a telescope be ground and polished as smoothly as possible and not deviate from their required tolerances by more than a fraction of a wavelength. With optical telescopes, that translates into only a few nanometers across the mirror's entire surface. This a challenging task, particularly during the final polishing of really large monolithic mirrors. It's less of a problem with segmented mirrors, where several smaller elements are combined and aligned individually to produce a composite mirror of much larger size. It is a lot easier to make small mirrors of excellent quality than to produce a large, optically perfect mirror surface. The flip side of course is that smaller mirrors must be kept in precise alignment at all times so as to maintain a combined surface accurate to a fraction of a wavelength.

An inherent problem with really large diameter mirrors is that they actually sag under their own weight. Since this

● The dome of the Gemini North telescope on Mauna Kea in Hawaii. To maintain ambient thermal equilibrium, the 46 m high and 36 m wide dome remains almost totally open when observations are underway.

distorts the mirror's optical surface, image quality will suffer unless corrective measures are taken. Another factor that affects image quality is mechanical flexure and stress in large instruments, which can lead to misalignment and poor tracking. To mitigate or overcome these problems, most modern telescopes are equipped with sensors that constantly monitor and correct both optical and mechanical changes. For example, "active optics" help correct primary mirror deformations generated by sagging and weight shifting. "Adaptive optics" minimizes image degradation caused by atmospheric turbulence, or "seeing," as astronomers usually call it. These technologies have launched a new era, particularly in solar astronomy where active and adaptive optics have become indispensable, especially for large telescopes.

OBSERVATORY DOMES

During the 1980s, extensive studies were carried out to evaluate the negative effects of atmospheric turbulence and seeing on image quality. The lessons learned have had a profound impact on the design of domes and the environment around telescopes. Prior to the 1980s, observatory domes were usually large hemispheric structures. They had narrow slits or openings for telescopes, which were completely isolated from their external environment. In contrast, modern observatory domes are as compact as possible, not necessarily round, and fully ventilated to minimize drafts and local air turbulence. ■

● The 1.2 m Swiss Leonard Euler telescope at La Silla Observatory. The spectrograph Coralie can be seen at the Nasmyth focus. This high-performance telescope is almost exclusively dedicated to the search for extrasolar planets.

■ When it was placed in service at La Palma Observatory in 1988, the 4.2 m William Herschel telescope was the third largest in the world. During this time exposure, the dome slit was kept open and rotated to show the interior of the structure.

La Palma Observatory

The immense volcano that once dominated the island of La Palma has long since gone silent. All that remains of the once proud cone and the lava cinders is the "cauldron," a huge, circular depression some 1,500 m deep and 8 km wide, known as the Taburiente caldera. Billowing clouds from the Atlantic Ocean touch and surround it from the southwest. This is the westernmost of the Canary Islands, and also the most jagged. Some 50 km long but only 30 km wide, it culminates with the 2,426 m high Roque de Los Muchacos. La Palma Observatory was built more than 20 years ago atop this small island on the largest volcanic caldera in the world. The 12 telescopes at the edge of the steep Taburiente caldera enjoy some of the finest skies on the planet and an average of 280 clear nights a year. Perhaps even more important, Spanish authorities have gone to great lengths to preserve this astronomical haven. This island of 80,000 inhabitants is not a major tourist attraction and strict guidelines are enforced to regulate public lighting and industrial pollutants, which ensures that La Palma remains a pristine location for astronomy. Not only has the summit of Roque de Los Muchacos been turned into a national park, but the entire island has also been designated a "worldwide astronomy preserve" by Spain and the independent government of the Canary Islands.

The Astrophysics Institute of the Canary Islands (AICI), comprised of more than 100 astronomers, manages the site, which is informally known as the Northern European Observatory. In late 2000, Spanish astronomers began construction of their own Grantecan telescope, which will be among the world's largest when completed in 2005. But before that they collaborated with many other European countries who were eager to install their most sophisticated equipment under the fine skies of Pico de Teide, Pico Veleta, Calar Alto and La

SPAIN

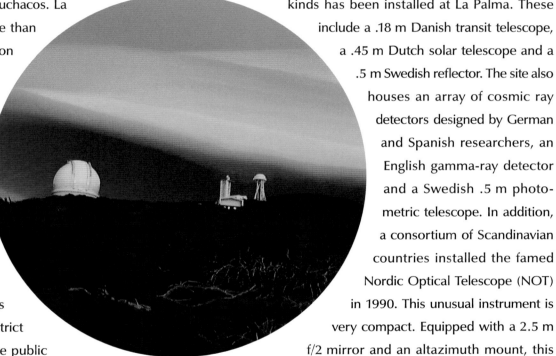

■ *Alongside Mauna Kea and Cerro Paranal, the La Palma Observatory ranks as one of the three most important astronomical sites in the world. The 10 m Grantecan will become operational at La Palma in 2005.*

Palma. Today, when visiting the observatory one finds not only Spanish astronomers but also researchers from England, Ireland, Italy, the Netherlands, Denmark, Sweden, Norway, Finland... It is rather surprising that despite its exceptional weather and viewing conditions, this site was only discovered by astronomers relatively recently. The first telescope was erected at the site in 1982, contrasting with 1966 for La Silla in Chile and 1970 for Mauna Kea in Hawaii.

Over the past 20 years, a panoply of telescopes of all kinds has been installed at La Palma. These include a .18 m Danish transit telescope, a .45 m Dutch solar telescope and a .5 m Swedish reflector. The site also houses an array of cosmic ray detectors designed by German and Spanish researchers, an English gamma-ray detector and a Swedish .5 m photometric telescope. In addition, a consortium of Scandinavian countries installed the famed Nordic Optical Telescope (NOT) in 1990. This unusual instrument is very compact. Equipped with a 2.5 m f/2 mirror and an altazimuth mount, this telescope weighs only 35 metric tons and cost less than US $13 million.

Britain and Holland are partners in the ING (Isaac Newton Group of Telescopes) project, which consists of three telescopes: the 1 m Jacobus Kapteyn, the 2.5 m Isaac Newton and the 4.2 m William Herschel reflectors. The Isaac Newton telescope, mounted in a classic 110-metric-ton equatorial mount, was initially sited at Herstmonceux, England, where unfortunately it averaged only about 50 good viewing nights a year. Once dismantled, modernized and brought online, it resumed operations at La Palma in 1984. For some time, the William Herschel telescope, also built in Great Britain, was the most powerful instrument at La Palma. With its huge altazimuth mount and 210 metric tons of moveable mass, this instrument was the world's third largest telescope when brought into service in 1988. This telescope features four

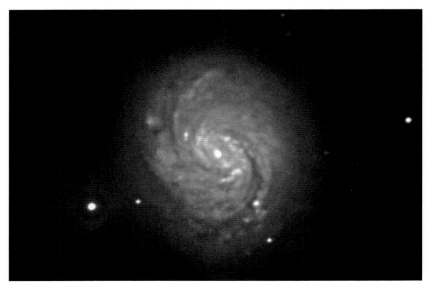

foci: a primary 11.8 m at f/2.8; a 46 m at f/11 Cassegrain focus and two f/11 Nasmyth foci.

The William Herschel telescope, equipped with adaptive optics, CCD cameras, infrared detectors and powerful spectrographs, is still one of the most productive instruments in the northern hemisphere. Among many accomplishments, it was first to reveal surface detail on another star (Betelgeuse), first to detect a gamma-ray burst event optically (GRB 970228), and it discovered the lowest temperature and hence the oldest white dwarf known to date (WD 0346+246). This star, located some 90 light-years from us, has a surface temperature of 3,500 K (3,227°C) and is probably more than 12 billion years old. The 4.2 m telescope has also produced the Herschel Deep Field Survey, one of the deepest cosmic surveys to date (see page 213). A total exposure time of 100 hours at the telescope's prime focus were required, using six different filters that covered the full spectrum from ultraviolet to infrared. Thousands of galaxies to magnitude 28 were recorded in a 7 x 7 arcmin field. Among other findings of cosmological significance, the Herschel Deep Field Survey recorded extremely faint red galaxies, with redshift probably exceeding z = 6. These objects are far too faint to be fully studied with only a 4.2 m telescope, and will probably be investigated in the future with the CMOS infrared spectrograph at the 8 m Gemini North telescope at Mauna Kea.

In 2000 a revolutionary new detector developed by the European Space Agency was installed at one of the William Herschel telescope's Nasmyth foci. The Super-Conducting Tunnel Junction Camera (S-Cam) is an extremely sensitive counter, capable of precisely measuring the exact arrival, wavelength and energy of each individual photon captured! The detection surface of the S-Cam, a tiny matrix of tantalum (a very hard and dense metal), is cooled to 1 K (–272°C) with liquid helium. This cutting-edge instrument with extremely fast measuring capabilities will observe such rapid phenomena as fast variable stars, eclipsing binaries, novae and pulsars.

■ *The 1 m Jacobus Kapteyn telescope took these images of spiral galaxies, with a CCD camera with a 10 x 10 arcmin field of view. The*

THE GIANT SPANISH TELESCOPE

Not far from the edge of the Taburiente caldera, Italian astronomers installed their 3.6 m national telescope in 1998. This altazimuth-mounted instrument is similar to the New Technology Telescope (NTT) at the European Southern Observatory at La Silla. In addition to several cameras and other receivers, this telescope is equipped with adaptive optics and a high-resolution spectrograph called SARG, with spectral resolution of 30,000 to 164,000, at wavelengths of 370 and 900 nm. It is also equipped for near-infrared work, thanks to a special Near Infrared Camera

true color of these galaxies is shown by combining images taken through three separate filters: blue, green and red.

Spectrometer (NICS) with a 1,000 x 1,000 pixel array, which covers a field of 4.2 x 4.2 arcmin at wavelengths ranging from 0.9 to 2.5 μm.

Spanish astronomers are contractually allocated 20 percent of observing time on all instruments at La Palma Observatory, since they provide infrastructure support, accommodations and their superb skies. Nevertheless, Spain, which has historically not had the funds for major astronomical facilities, decided in 1996 to embark on a truly giant telescope venture. Called the Grantecan, for *Gran Telescopo Canarias*, it will be a 10 m telescope modeled after the twin Keck telescopes at Mauna Kea. Costing an estimated US $123 million, the facility will receive 95 percent of its funds from Spain and 5 percent from Mexico. Grantecan will be supported by a 300-metric-ton altazimuth mount and will be equipped with a segmented primary mirror consisting of 36 Zerodur hexagonal elements produced by the French firm Reosc. The segmented primary mirror will be computer-controlled and adjusted through servomotors and actuators. Since Grantecan was built in 2002 and will be placed in service in 2005, a dozen years after Keck I, it has benefited greatly from recent advances in optical electronics. This giant Spanish telescope is expected to perform markedly better than the two American instruments. In addition to CCD cameras,

spectrometers and adaptive optics, Grantecan will also be equipped with a particularly advanced receiver, Canari-Cam, a device likely to produce results similar to those expected of larger telescopes in the future. With this infrared imaging spectrometer, sensitive between 8 and 24 μm, Grantecan's large aperture should provide angular resolution near 0.2 arcsec at 8 μm and 0.5 arcsec at 24 μm.

The center of our galaxy will be one of the main targets of the 10 m La Palma giant. Canari-Cam should be particularly effective for high-resolution infrared imaging of regions in and around the giant black hole at the core of the Milky Way. At 8 μm for example, Grantecan should resolve detail of the order of 500 AU. One AU is equal to the distance between the Earth and the Sun, or approximately 150 million km. Spanish and Mexican astronomers, however, also hope to *directly* image extrasolar planets with Grantecan and Canari-Cam, planets that have so far been detected by indirect means only. At a distance of 50 light-years, the 10 m telescope should be able to resolve objects separated by only 2 AU. That's because giant gas planets, like Saturn and Jupiter, and larger objects radiate the strongest in the infrared, but the brightness of their parent stars is significantly attenuated. It's quite likely that by the end of this decade, Grantecan will have imaged several exoplanets and begun to study their physical properties directly. ■

The Jodrell Bank Radio Observatory

Bathed in sunlight and with a gentle sea breeze wafting from the direction of Ireland, dairy cows graze happily in the soft green meadow. This tranquil corner of the English countryside, like a scene from a Cheshire postcard, contrasts sharply with industrialized Manchester some 30 km away. Seemingly motionless and pointing to a small spot in the sky, the great dish of Jodrell Bank is visible from several kilometers away in the bright sunlight. Almost imperceptibly, the gigantic antenna turns westward arc second by arc second, tracking the pulsar PSR 0329+54 as it nears the horizon.

The dream of one individual, Sir Bernard Lovell, this great radio telescope has been a dominant feature of the rolling Cheshire countryside for over 40 years. After World War II, Lovell took advantage of the new radar technology, which had been developed in Britain against German air raids. After first experimenting with rudimentary radar antennas to locate incoming meteors, Lovell and engineer Charles Husband began plans in 1950 for a giant radio telescope that, like ordinary optical telescopes, would be able to point to and follow any object in the sky.

UNITED KINGDOM

■ *Several weeks after its installation, Lovell's radio telescope became world famous when it followed Sputnik, the world's first artificial satellite, which was launched October 4, 1957.*

The two men faced major technical and financial difficulties, however, and when their 76 m radio telescope dish entered service seven years later, its cost was ten times greater than originally estimated! It must be stressed that radio astronomy was still in its infancy at the time, and the project required many modifications and entirely new applications even while under construction. To keep costs down, the telescope was initially intended to observe in the 1 to 10 m wavelength range. That required only a lightweight antenna with a metal mesh surface accurate to just a few tens of centimeters.

However, by 1951 a number of researchers had discovered that neutral hydrogen, so abundant in nebulae and galaxies, emits strongly at 21 cm, prompting Lovell to modify his giant telescope accordingy. This made it possible several years later to reveal the spiral structure of our Milky Way galaxy and to detect galaxies at the edge of the visible Universe. Thus a radical design change was required partway through the telescope's construction. Rather than a mesh surface, the antenna now had to be covered with more than 7,000 metal panels, each 2 mm thick and perfectly polished to provide a surface accurate to only a few centimeters.

The Jodrell Bank telescope was dubbed Mark I when first placed in service, but renamed the "Lovell Telescope" in 1987. Its development had a tremendous impact on the astronomical community, as well as the public, who had never before seen a scientific instrument of this size. The telescope is nearly 90 m high and its total movable weight approaches 3,200 metric tons. It rotates on a circular track more than 110 m in diameter. The 76 m antenna has a surface area of 4,560 square meters and weighs 1,500 metric tons.

Even today there are only two other radio dishes in the world that are larger than Jodrell Bank: the 100 m dish erected at Effelsberg, Germany, in 1972, and the 100 m antenna at Green Bank, United States, placed in service in 2001. During its 40 years of operation, the Lovell Telescope has undergone several mechanical and electronic upgrades and improvements. The antenna has been completely resurfaced with new reflective panels, and outfitted with receivers thirty times more sensitive than the originals. The most recent resurfacing in 2002 has increased resolution to .06 m.

During its 40 years of exploration, this venerable radio telescope has been particularly productive in research on pulsars, those mysterious objects accidentally discovered in 1967 by Anthony Hewish and Jocelyn Bell with the Cambridge

radio interferometer. These invisible objects, whose discovery earned Hewish the Nobel Prize for physics in 1974, are powerful radio sources that emit regular pulses approximately once per second. They were studied in detail at Jodrell Bank and were soon identified as lying at the core of dead stars. Dozens of pulsars were subsequently discovered at various points in the sky, usually associated with remnants of exploded supernovae. Rotating at extreme speeds, up to a thousand times per second, pulsars act like giant dynamos that emit bursts of gamma and X-rays from their magnetic poles.

MORE THAN A THOUSAND PULSARS

As veritable atomic stars, pulsars contain more mass than the Sun but are only about a dozen kilometers in diameter. Also known as "neutron stars," they are perfectly spherical bodies comprised almost entirely

■ *The Lovell radio telescope has undergone many upgrades and improvements since 1957. In 2002, the 4,560 square meters was resurfaced with new, more accurately polished panels. The telescope's drive mechanism, consisting of six powerful motors, was also modernized.*

■ *The control room at Jodrell Bank Observatory. Lovell Telescope operators work around the clock to follow and record radio signals from pulsars and quasars.*

of a hyper-dense fluid of neutrons, a cubic centimeter of which would weigh a billion metric tons. At least 1,000 pulsars have been discovered so far. They are not readily detected at optical wavelengths and about two-thirds have been discovered jointly by Jodrell Bank and their colleagues at Parkes Observatory in Australia. Although Jodrell Bank has the capacity to detect faint and distant pulsar signals, it cannot locate these objects in the sky with great precision. Working at an .18 m wavelength, the Lovell Telescope's resolving power is limited to about 10 arcmin, or one-third the apparent diameter of the Moon. Despite its impressive 76 m diameter, the Jodrell Bank antenna is somewhat myopic, like all radio telescopes. To overcome this and increase the resolving capabilities of their single dish, British astronomers linked several different antennas into an immense radio interferometer array called Merlin. ■

The Merlin Interferometric Array

UNITED KINGDOM

The radio telescope at Jodrell Bank is seemingly fixed on the pulsar PSR 0329+54. In an effort to improve its acuity nearly a thousandfold, the giant dish's complex observing run has gone on for several hours. However, Jodrell Bank is not alone in this effort. Several distant partners well beyond the Cheshire countryside, much farther south and east, have joined it. This is a pivotal moment in radio astronomy, since it is the first observing run of Merlin, the Multi-Element Radio Linked Interferometer Network in Britain, established as a gigantic telescope array extending more than 200 km between Jodrell Bank, Knockin, Defford and Cambridge. Merlin is composed of seven antennas in total, including five 25 m diameter dishes, the 32 m at Cambridge and of course the 76 m antenna at Jodrell Bank. To obtain really high-resolution images, the interferometer must operate for several hours and progressively integrate signals from pairs of antennas (21 combinations in all) in steady cycles to encompass the entire Merlin array.

The Merlin interferometer became operational in 1980 and, along with the Very Large Array (VLA) in the United States, is one of the most powerful instruments of this type in the world. Covering the .01 m to 1 m wavelength range, Merlin can resolve down to an impressive 0.05 arcsec at 6 cm, which is comparable to what the Hubble Space Telescope can achieve in the visible range.

Like Jodrell Bank at its core, the Merlin array is managed by the University of Manchester and is in great demand by British astronomers doing research on pulsars. In fact, over the past 20 years, the very high resolving power of this interferometer has helped pinpoint the exact location of all the pulsars the 76 m Jodrell Bank dish discovered during its original systematic sky survey. Merlin has been particularly useful in tracking the motion

■ *The Jodrell Bank radio telescope is the core element of the Merlin interferometer. Six other antennas ranging from 25 to 36 m in diameter form the rest of the array. Merlin is comparable to similar arrays including the VLBI, VLA and VLBA.*

of the pulsar within our galaxy. Such measurements have uncovered a truly remarkable fact; almost half of the pulsars observed move so fast that they will escape the gravitational influence of the Milky Way in a few million years and escape into the intergalactic realm.

To get a better understanding of how supernova explosions affect the interstellar medium, Merlin began a systematic search for supernovae in M82, a galaxy in Ursa Major some 10 light-years from our own. This was seen as a particularly interesting target for radio astronomers. M82 is an extremely active galaxy in which several hundred new stars are born annually, and which is expected to contain an unusually high number of supernovae. Throughout more than a century of surveillance, no explosions had been observed. This was most likely due to this galaxy's exceptionally high gas and dust content, which effectively masked any evidence of star birth or death. In the 1990s, however, working at a wavelength of 6 cm, Merlin discovered 35 supernovae remnants or shells, all of which had probably exploded within the last thousand years, including as recently as 30 years ago.

MERLIN JOINS VLBI AND VLBA

Despite having an effective diameter of 200 km, Merlin was still not powerful enough to resolve really fine detail in the residue of stellar explosions. In order to achieve this, astronomers created an "array of arrays" of radio telescopes, by combining Merlin, the VLA in the United States and the European Very Long Baseline Interferometry (VLBI) array. The VLBI, currently linked with Merlin, consists of radio telescopes located in Finland, Sweden, Poland, the Ukraine, the Netherlands, Germany, Italy and Spain. This multinational effort made it possible to observe M82 with a virtual telescope some 12,000 km in diameter, providing the

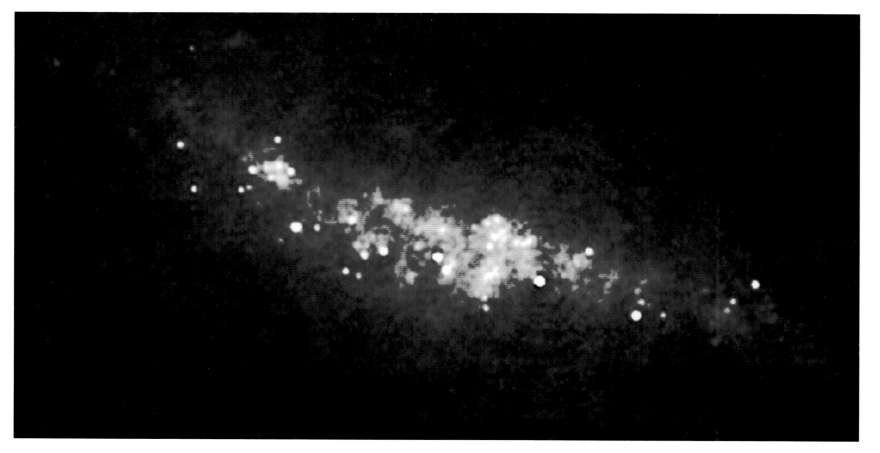

■ *This image of M82, taken at a wavelength of 6 cm, combines data obtained by the Merlin and VLA radio interferometers. The bright round spots are the supernova gas remnants. The field of view of this image is about 60 arcsec.*

unprecedented resolution of 0.001 arcsec (one thousandth of an arc second!). At the distance of M82, this corresponds to detail only a few light-months across. In this way, by comparing images of the youngest supernovae in M82 obtained over several years, researchers were able to determine that the resultant gas shells expanded at the extraordinarily high rate of more than 20,000 km per second, or close to 10 percent the speed of light.

M82 is one of the closest galaxies known, but Merlin is called upon to study objects at cosmological distances that are not just millions of light-years away but billions. The Merlin array has been particularly effective for systematic investigations of gravitational lensing phenomena. Among the first to be studied was the double quasar Q 0957+561 in Ursa Major, discovered in 1979 by the Jodrell Bank radio telescope. Gravitational lensing effects are caused by very large objects such as galaxies or galaxy clusters, which happen to lie directly in our line of sight of more distant galaxies or quasars. The gravitational field of the closer object acts like a giant lens that bends and magnifies the light from the more distant source, often producing multiple images of it.

Current research efforts use high-resolution gravitational lensing images

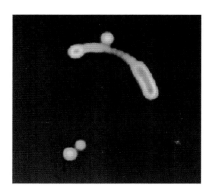

■ *Gravitational lensing effect images of 1938+666, as recorded by the Hubble Space Telescope (upper) and by Merlin (lower). The Merlin image is twice as clear and shows the effect in greater detail. The galaxy causing the gravitational lensing is not visible.*

obtained by Merlin to help clarify cosmological space/time characteristics. For example, when gravitational lensing produces two images of a distant quasar, the radiation reaching Merlin has followed different paths around the lensing object. Should the quasar undergo a change in brightness, this will appear at different times for each image, because gravitational lensing has deflected the light from each image differently. Such delays can amount to several months or even years. Once the exact trajectory of the radiation from each image is determined, a relatively simple calculation can be applied to estimate the Hubble constant, or the rate of expansion of the Universe. This calculation will incorporate such parameters as the apparent distance of the lensed object, the spectral characteristics of the gravitational lens and the time differential involved. This work is of fundamental importance. Recent estimates place the Hubble constant at approximately 75 km/s/Mpc, based on expansion rates of comparatively close objects measured by the Hubble Space Telescope. By probing the Universe at much greater distances through gravitational lensing, Merlin can measure expansion rates out to six or eight billion light-years and determine if they have changed in the course of time. ■

> Radio astronomy: observing the invisible

*E*ver since the discovery in the 1930s that natural radio waves emanated from space, radio astronomy has been an indispensable tool in our study of the Universe. Despite its relatively recent development, radio astronomy has made several fundamental discoveries. Whether it involves individual dishes or huge arrays, the tools of radio astronomy are advancing rapidly, driven by spectacular results and new discoveries at every turn.

THE BREADTH OF THE RADIO WAVELENGTH DOMAIN

The radio domain encompasses the largest portion of the electromagnetic spectrum, extending from the submillimeter range to decameters. This covers a broad array of signals originating from both thermal and non-thermal sources, including magnetic and synchrotron radiation, which is produced by high-speed electrons spiraling through magnetic fields. The radio spectrum also includes several distinct atomic and molecular signatures, like the .21 m radiation from hydrogen. Lastly, thanks to the Doppler effect, radio observations at very long wavelengths have detected some of the most distant objects in the Universe.

The radio end of the spectrum is replete with new, often serendipitous discoveries. For example, pulsars and quasars were first discovered through radio astronomy, as was the cosmic microwave background radiation at 2.7 K, the residue of the very high temperature of the very early, extremely dense Universe.

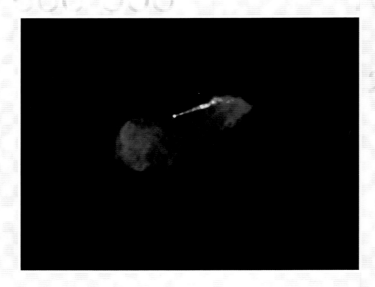

● The galaxy M87 in Virgo is about 50 million light-years from us. Two jets emanating from its nucleus are evident in this 20 cm image obtained with the VLA interferometer.

GIANT RECEIVERS

Astronomical observations at radio wavelengths require large-aperture receivers for two main reasons. First, most celestial radio signals are extremely weak and, second, large-aperture telescopes are needed for sufficient angular resolution at the relatively long wavelengths involved. Consequently 30 to 100 m in diameter dishes have been erected at Pico Veleta, Nobeyama, Parkes, Effelsberg and Green Bank, among other locations, in order to cover the millimeter, centimeter and decameter ranges of the spectrum.

Despite their impressive size, however, those facilities are still not large enough to provide resolution comparable to that obtainable at shorter wavelengths. The diffraction limit of a telescope is proportional to the wavelength of the radiation it receives and inversely proportional to the diameter of its collecting surface. Consequently a radio dish working at 2 mm must be a thousand times larger than a telescope working at 2 μm to attain comparable angular resolution! For example, to resolve detail comparable to an 8 m optical telescope with adaptive optics working at 2 μm, a radio telescope observing at 2 mm would have to be 8 km in diameter, and one working at .2 m, 800 km!

INTERFEROMETERS

Since it is clearly impossible to build single antennas several tens or even hundreds of kilometers in diameter, radio astronomers were quick to adopt interferometry as an alternative technique. When two telescopes observe concurrently and are physically separated by a defined distance or baseline, their effective angular resolution is equal to that of a single dish of the same diameter as the baseline. In practice, radio interferometry is considerably easier to achieve than optical interferometry, because the constraints of this technique are much greater at optical wavelengths.

The first radio-interferometry arrays were developed in the 1960s and used for simultaneous observations of single targets. These developments led to the famous facilities at VLA some 20 years later. To attain even greater angular resolution, interferometry is now carried out at an even larger scale, involving groups of radio telescopes separated by thousands of kilometers and sometimes located on separate continents. An example of this is VLBI or Very Long Baseline Interferometry. Unlike classical interferometers, these arrays are not linked directly by cables or other means with their signals recombined immediately. Instead, each telescope records data independently but is precisely timed by atomic clocks. The signals are later synchronized and combined. In the United States the VLBA has been in use since 1992. It consists of some 10 antennas covering the northern hemisphere from Hawaii in the Pacific across to the east coast of the continental United States. European countries have put together a core array of 10 radio telescopes called JIVE or Joint Institute for VLBI in Europe. Facilities like these can attain angular resolution of a few thousands of an arc second in the centimeter wavelength range and some 50 millions of an arc second at millimeter wavelengths!

RADIO TELESCOPES IN SPACE

Not all radio waves traverse our atmosphere freely; anything shorter than 1 mm is mostly absorbed, while waves longer than 15 m are reflected back by the ionosphere. Submillimeter wavelengths can be observed at extremely arid, high elevations like Antarctica and some of the highest mountain locations. Because of this, submillimeter receivers functioning singly or as interferometers have been installed on Mauna Kea in Hawaii, since it rises above 97 percent of atmospheric water vapor. Unfortunately this site is not ideally suited for more detailed observations at these wavelengths.

Because of these limitations, radio telescopes have been sent aloft on balloons and space observatories, often on missions dedicated to very specific areas of research. That was the case with the orbiting Cosmic Background Explorer (COBE), which took accurate measurements of the cosmic background radiation in 1989–90. Its successor, the much-anticipated Planck orbiting observatory, should obtain even more precise measurements a few years from now.

Another advantage of going into space is that truly gigantic interferometer arrays could be launched and attain levels of

● Just like regular optical telescopes, some radio telescopes are completely movable and can be pointed in any direction. The largest fully steerable radio dish in the world, at Green Bank, West Virginia, is 100 m in diameter. A 30 m dish is located at Pico Veleta (top). Much larger instruments, like the 300 m dish at Arecibo or the 200 x 35 m Nancay radio telescope shown here (middle), are only partially movable. Finally, radio interferometers consisting of arrays of moderate-size dishes can be combined to function as virtual telescopes of much larger size, from tens to thousands of kilometers across. Pictured here (bottom) is the VLA in New Mexico, which consists of 27 dishes, each 25 m in diameter.

resolution superior to even the longest baselines on Earth. Several exploratory missions along these lines have been planned, including a joint Japan–United States effort called VSOP or VLBI Space Observatory Program. The Japanese HALCA satellite with an 8 m antenna was placed in a high permanent orbit in 1997 and formed baselines with ground-based telescopes up to three times longer than the VLBI. This is probably just the beginning, however, as one day a Moon-Earth baseline will probably be established. ■

The Effelsberg Radio Telescope

I n Westphalia, about 50 km from Bonn and Cologne, the atmosphere is often laden with clouds and torrential rain. Only barely illuminated by the car's headlights, the narrow road winds through the night, amid dense fog and cold rain. Suddenly, a ghostly figure appears ahead of us, illuminated by a bolt of lighting and announced by a clap of thunder. The giant Effelsberg radio telescope slowly turns a few degrees toward some invisible celestial object. With an instrument capable of cutting through rain, fog and clouds, German astronomers pay scant attention to the inclement weather. Perched in his control booth 100 m down the hill, a technician regulates all operations at this huge facility. As row upon row of monotonous numbers scroll across the computer monitor, the drive motors automatically move and point the telescope at any target. Like Ulysses' famed Cyclops, Polyphemus, it points to but cannot really see its very distant target, the quasar QSO 1928+73.

GERMANY

Effelsberg is among the most powerful, fully mobile radio telescopes in the world, second only to the Green Bank radio telescope in West Virginia. Built in 1972, Effelsberg is a very impressive instrument. It has a 100 m diameter dish, weighs 3,200 metric tons and is 110 m high.

■ Four support arms hold the secondary receiver at the prime focus of the Effelsberg radio telescope. This secondary receiver reflects part of the radio signals toward the instrument's secondary focus. A primary observing cage is located behind the secondary receiver.

This giant instrument was designed to observe the sky at wavelengths ranging from 3.5 mm to .9 m. Like traditional optical telescopes, it is set in an altazimuth mount, and despite its massive size, it can track with a precision of just a few arc seconds. The telescope's azimuth rail is 64 m in diameter, and the half-wheel that drives it in latitude has a radius of 56 m! The whole assembly is remote-controlled from a nearby command module. Working from a 4 x 8 m bay window, the operators control all aspects of the facility, including the telescope, its multiple detectors and the near one-hectare of lawn underneath it. This particular area is off

limits during all observing sessions to ensure that no one leaves a wrench by mistake near one of the many metal supports so integral to the telescope. Most of the operators working at this large facility are former sailors who are used to moving high-inertia masses such as container barges or oil liners. They take their jobs very seriously, always mindful of the accident at the 91 m Green Bank radio telescope, which came crashing down in the middle of the night.

There are no actual astronomical images on the control room computer screens, just rows and rows of coordinates rapidly changing as the telescope moves and observes. The weak radio signals are built up gradually, and astronomers will use them later to generate computer images of the objects observed.

Few astronomers actually visit and use this large instrument firsthand. Instead, they submit their observing programs directly to the Effelsberg engineers and technicians, who input them and collect data 24 hours a day, year round.

The technicians who maintain the telescope must ascend 50 m by elevator to a platform where they can access the receiver assembly. The telescope is pointed straight at the zenith at such times so that technicians, clearly unafraid of heights, can climb along one of the secondary support arms. This long tube, made of a metal trellis, bears an uncanny resemblance to some of the images created by the artist Escher. It consists of both a staircase and a ladder set at 90-degree angle to each other, which provide both an entrance to and an exit from the instrument in any position.

A TECHNOLOGICAL TOUR DE FORCE

The observation cage is located at the very top of the telescope. It is a metal cylinder, roughly the size of one of the modules of the

International Space Station, and holds about a half-dozen radio receivers. Accessed via a metal portal, the cage is quite spacious and gives the distinct impression of being in a space station. This is further reinforced by the presence of a staircase and metal gripping bars located on all horizontal and vertical surfaces. The cage glides slowly in the wind and vibrates silently to the rhythm of the cooling pumps.

Relatively speaking, the surface of the 100 m antenna, which consists of 2,400 aluminum panels, is as precise as that of any optical telescope. However, while a 10 m optical mirror must work in the 500 nm wavelength range, the Effelsberg antenna, designed to observe in the centimeter and millimeter range, can do with a surface accuracy of only 0.5 mm. The technological challenge for a dish of this size is to maintain such precision over its entire surface area of 7,850 square meters, regardless of its orientation or any fluctuations due to high winds.

■ *Built in 1972 under the direction of the Max Planck Institute for Radio Astronomy, the 100 m Effelsberg radio telescope is one of the key elements of the European Very Long Baseline Interferometry (VLBI) array. This remarkably sensitive instrument is used primarily for studies of pulsars, galaxies and quasars.*

Despite its imposing size, the Effelsberg radio telescope can attain only limited angular resolution, namely 9 arcmin at 21 cm and 1 arcmin at a wavelength of 3 cm (approximately the resolving power of the naked eye), and 10 arcsec at 3.5 mm (about the resolution of Galileo's telescope). This is why, for at least a third of its observing time, this radio dish is linked to the array of European radio telescopes known as the Very Long Baseline Interferometer (VLBI). Through this virtual radio telescope, which is several thousands of kilometers in diameter, Effelsberg can attain angular resolutions between 0.01 and 0.001 arcsec, as is required for research in modern astrophysics. On the other hand, the Effelsberg radio telescope has exceptional sensitivity and can detect radio signals as weak as 10^{-18} watts. Because of this, German radio astronomers have become specialists in the study of the farthest galaxies and quasars in the Universe. ■

The Virgo Gravity Wave Telescope

We are in a small, very flat corner of Tuscany not far from the village of Cascina. The sun-drenched countryside is punctuated here and there by tall Italian cypress trees, and crisscrossed by the meandering Arno River. In the distance, you can just make out the highway from Pisa to Florence, where long ago Galileo spent many a day meditating about the Universe and many a night observing the stars. Driving along this isolated country road, you are completely unaware of the presence nearby of one of the strangest scientific instruments ever designed, the mammoth Virgo gravity wave telescope. To fully appreciate the size of this interferometer it must really be seen from the air, where it looks like a perfect "L," seemingly suspended above ground and with two long branches stretching out forever.

This telescope was obviously not located in Tuscany because of the quality and clarity of the Italian skies. At an elevation of just 5 m, the site is often subjected to cold fog from the nearby Mediterranean, which completely envelops the interferometer's metal scaffolding. Virgo is not bothered by such inclement viewing

ITALY

■ This is an inside view of one of the two arms of the Virgo gravity interferometer as it was being built. The observatory is located near Pisa, one of the most seismically stable regions in Tuscany.

conditions because it does not measure any type of electromagnetic radiation. This extraordinary instrument was designed by physicists in order to prove that Einstein is correct and who believe that this work will open another "window" to the Universe.

Albert Einstein predicted gravitational waves back in 1916 when he published his theory of general relativity. According to this theory, the universal space/time continuum is dynamic and curved under the influence of large mass/energy objects like stars, galaxies or clusters of galaxies. The space/time "surface" is generally calm and smooth. However, some of the most violent and energetic events in the Universe, including binary pulsars, black holes, supernovae and hypernovae, can generate

gravitational waves, which dissipate at the speed of light and deform the space/time continuum. Gravitational waves are like the ripples in a smooth pond caused by individual drops of rain.

Virgo was designed to detect space/time ripples caused by cosmic events of such magnitude that they surpass anything on a human scale. Some scientists believe that they already have indirect evidence of such waves. American astronomers Joseph Taylor and Russell Hulse have monitored the binary pulsar PSR 1913+16 at Arecibo observatory for several years, and have noted a progressive slowdown in the pair's rotational period. This effect is attributed to enormous dissipation of energy in the form of gravitational radiation. This discovery, one of the most elegant in modern physics, earned them the Nobel Prize in Physics for 1993. As they lose energy, the two neutron stars in PSR 1913+16 are spinning toward each other at about 2 m per year. In less than 300 million years, they will fuse violently and release an enormous amount of gravitational energy. The Virgo telescope in this serene Tuscan setting was built specifically to observe such cataclysmic events.

While recognized as a cutting-edge facility, Virgo is actually modeled after a classic instrument in physics, the Michelson interferometer. The core of the gravity wave telescope is a powerful and extremely stable laser, whose beam is split by a semireflective blade. Half of the light passes through the blade and continues into one of the instrument's 3 km long arms and the other half is deflected at right angles into the other arm. The two beams are reflected back to the center by mirrors at the back of each tunnel, and recombined to form an interference pattern. According to the Italian and French designers of the telescope, when gravitational waves enter the interferometer, the slight alterations in the space/time continuum will disrupt the two perpendicular laser beams, resulting in an immediate dephasing of Virgo's interference pattern.

Although the principles underlying this experiment are quite simple, the instrument itself is one of the most complex and precise ever built. That's because the anticipated space/time alterations generated by gravitational radiation are infinitesi-

■ *Virgo is a Michelson-type interferometer designed to observe gravitational radiation emitted by binary pulsars, supernovae and hypernovae. Physicists and astronomers are working jointly on this innovative project.*

mally small. It is estimated that when a single gravitational wave passes through it, one of Virgo's arms is expected to shorten by only 1 billionth of a millionth of a millimeter! This level of precision is at the very limit of current technology, which is why Virgo is truly an adventure in metrology for the engineers and technicians who built it.

WHEN VIRGO OBSERVES VIRGO

In order to increase the interferometer's sensitivity, both laser beams are sent through a series of reflective mirrors in each arm of the instrument, called the "Fabry-Pérot Cavity." This subjects each beam to 40 internal reflections before it reaches Virgo's focal point, and extends the effective length of each arm to 120 km. Each arm is actually a long tube maintained under high vacuum. All mirrors are polished to an accuracy of a few nanometers and suspended from a series of gas-filled springs, so as to totally dampen all vibrations generated by wind, nearby cars or any minor seismic activity.

Virgo became operational in early 2003. Although constantly subject to interference by Earth-generated background noise and "breathing" cycles, the instrument is expected to detect the passage of gravitational waves from time to time. Astronomers have pinned their hopes on the telescope's namesake, Virgo, one of the nearest and largest clusters of galaxies in the constellation Virgo. This cluster contains

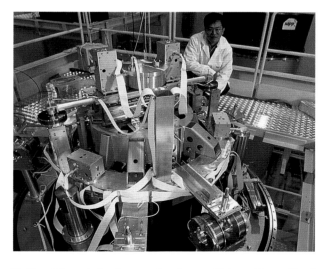

■ *The Virgo interferometer is a collaborative effort involving several laboratories of the French National Council for Scientific Research (CNRS) and the Italian National Institute of Nuclear Physics.*

several billions of millions of stars. It is expected that on average, one supernova per month will explode there and that a couple of neutron stars will fuse together there about once a year. Hence it would not be unexpected if the first gravitational wave recorded came from the direction of the constellation Virgo. From a strict scientific standpoint, however, such an event would not be proof positive of Einstein's prediction, because this first-ever gravity wave telescope is rather myopic. How, for instance, would we differentiate between a mere seismic "sigh" and a true gravitational wave? Ideally the arrival of such a wave would coincide with something that could be verified with regular telescopes: for example, the discovery of a supernova in the Virgo cluster. This would still be challenging, however, since the resolving power of the Virgo telescope is very poor. Ironically, while gravity waves can easily penetrate all forms of matter in the Universe, including galaxies, stars and planets, light rays are often absorbed by interstellar gases and dust, causing most supernovae to go undetected.

The cure for Virgo's myopic condition is likely to come through linkages with facilities such as "Ligo," two similar gravity wave telescopes placed into service in 2003 in the United States. The three interferometers will function as an array, with a collective separation of several thousands of kilometers. Should they detect a gravitational wave simultaneously, astronomers will be able to pinpoint its source to within a few degrees.

Such cataclysmic events as the fusion of binary pulsars, the formation of black holes and violent stellar explosions will be of major interest to the scientists who built and operate Virgo. But, who knows what else this opening of a new window to the Universe might reveal? No doubt, it will be something completely unexpected. ■

■ *The Zelentchouk telescope is the largest astronomical machine ever built. Including its three-story-deep drive mechanism, the instrument is nearly 45 m high. Despite its total weight of more than 650 metric tons, however, the entire unit is easily moved by hand.*

The Zelentchouk Observatory

The Russian cosmologist Igor Karachentsev closes the elevator door behind him. Half a minute later and 15 m higher, he enters a lit room, packed with racks of electronic equipment and furnished with a desk and some chairs, where his colleagues await him. At the end of this room is a stairway leading farther into this unusual structure. The three men adjust some dials and turn on a TV monitor showing several stars shimmering under a turbulent atmosphere. Karachentsev and his colleagues are getting ready to observe the sky from inside the largest telescope ever built! The BTA-6 (Bolchoi Teleskop Azimoutalnyi-6 m) is truly a telescope of superlatives: housed in a 50 m high dome, its tube is nearly 30 m long and its single mirror is 6 m in diameter. The telescope's drive mechanism is located in the basement, some 15 m below ground. This includes a 6 m diameter worm gear and six hydraulic levels floating on a film of oil under 40 bars of pressure, which support the entire 650-metric-ton instrument.

RUSSIA

■ *The giant 6 m Zelentchouk telescope was completed in 1976. It was the largest telescope in the world for about 15 years, but today ranks only 18ᵗʰ in size.*

Working at one of the three foci of this 6 m telescope, which include the 24 m prime focus and two 180 m Nasmyth foci, the three Russian astronomers are not even aware of the instrument's drive motor. The control center is housed in one of the vertical pillars of the fork mount, and the observers are barely aware of their slow movement in the azimuth.

This colossal steel structure is the creation of astronomer and engineer, Bagrat Ioannissiani. When this ambitious project was planned in 1960, Ioannissiani suggested what was then a revolutionary notion: to equip this giant telescope with an altazimuth mount. This type of mounting offered several advantages over the traditional equatorial mounts used by all large telescopes at the time: stability, simplicity and almost no flexures. The down side of this type of mounting is that it requires sophisticated and powerful computers to operate effectively: something that the

Eastern bloc astronomers did not have until 1980. It also took a very long time to render BTA-6 operational, from its conception in 1960, to the start of construction in 1965 and to first light in 1976. At that time, BTA-6 overtook the 5 m Hale telescope at Mount Palomar as the world's largest, a title it held until 1991, when the 10 m Keck telescope was placed in service on Mauna Kea.

Though a distinct technological success, the BTA-6 telescope never produced the high-quality scientific results that the Soviet and later the Russian researchers hoped for. In retrospect, both the researchers and the observatory were at a disadvantage from the very beginning, located in the remote community of Zelentchouk. First, in 1960, the central Soviet regime in Moscow decided to locate the observatory atop Mount Pastoukhov in the Caucasus, where climatic conditions are far from ideal. Secondly, the observatory dome was poorly designed and too large, and often contributed to the atmospheric turbulence that meant the telescope rarely attained resolution better than 2 arcsec. Lastly, just as the giant instrument was placed in service, western astronomers began using the first CCD cameras and literally doubled the performance of their telescopes. Since Russian astronomers did not have access at the time to this new image capture technology, the 6 m telescope could never compete with its counterparts in California, Hawaii and Chile.

Nonetheless, Bagrat Ioannissiani remains a notable figure in the annals of astronomy. The novelty and foresight of his project earned the respect and admiration of astronomers worldwide, many of whom came to use this huge instrument alongside their Russian colleagues from time to time. In the dim light of dusk, the imposing dome of this great telescope seems like a cathedral turning slowly toward our galaxy and creating the illusion that you can touch the stars. ■

> Atmospheric absorption and turbulence

In many ways our Earth's atmosphere is a source of inspiration and beauty for poets and casual observers alike: the blueness of the daytime sky, the ruddy colors at sunset and the twinkling stars on a summer night. Our atmosphere also protects us from the harmful ultraviolet rays of the Sun. For astronomers, however, this thin layer of gases is a source of many problems, including the absorption, refraction and diffusion of light from celestial sources, as well atmospheric glow and turbulence, all contributing to the degradation of the images they want to observe.

● Atmospheric refraction is familiar to all who have observed a sunset or sunrise: the Sun's disk appears flattened at the poles, no longer as circular as it does when high up in the sky. Often under these circumstances a flash of green light is momentarily visible at the top of the solar disk, because the shorter wavelengths are refracted less strongly than the longer wavelength rays.

ATMOSPHERIC FILTERING

Many of the atoms, ions and molecules in our atmosphere act like strong filters and effectively block out whole sections of the electromagnetic spectrum. Consequently our "window" through the atmosphere is open only to those wavelengths that can traverse it freely or are only partially attenuated in the process. At sea level, for example, this is limited to the optical range, infrared and radio waves. Since absorption by the atmosphere is greatest at sea level and declines with altitude, really high elevation locations on Earth (like Mauna Kea in

Hawaii) are prized astronomical observing sites. Locations like Cerro Paranal in the Atacama Desert of Chile, though "only" 2,500 m high, are also favored because of their extremely arid climate and low levels of atmospheric water vapor.

The rest of the electromagnetic spectrum, including gamma and X-rays, and portions of the infrared and radio spectrum, is only accessible from space. Several approaches have been used to achieve this, including airborne observatories on specially modified jetliners, high-altitude balloons and, of course, satellite observatories orbiting the Earth.

All atmospheric components, be they gas, dust, particulates or aerosols, diffuse and scatter visible light and other wavelengths to varying degrees. That's why the sky appears blue during the day and reddish at sunset. Only radio waves are not affected much by diffusion, which is why radio astronomy can be done around the clock. Moonlight and artificial light from cities and other urban centers are also scattered across the night sky and affect astronomical observing. That is why most astronomical observing sites are located as far away as possible from urban light pollution, and studies of really dim celestial targets are confined to times before and after the new Moon.

ATMOSPHERIC GLOW AND FLUORESCENCE

Another important atmospheric phenomenon affecting astronomical observations is sky glow. This is caused by "fluorescence," or a release of photons due to interactions between electrons and ions in the atmosphere. Such emissions may be "continuous" in nature, and affect a wide range of wavelengths, or they may be short bursts affecting selected portions of the spectrum only. Sky glow can be intense enough to significantly dim faint objects like galaxies and nebulae. To overcome this, astronomers must monitor and quantify the intensity of sky glow and then subtract this from their data and measurements.

THERMAL EMISSIONS

The atmosphere is also a source of heat, which is manifested as infrared radiation and at radio millimeter wavelengths. Thermal emissions in our atmosphere can be strong enough to completely mask signals at those wavelengths from astronomical sources. Atmospheric emissions must be subtracted, therefore, from the astronomical signals by taking alternate readings of the object and an adjacent portion of the sky.

ATMOSPHERIC REFRACTION AND DISPERSION

When light rays pass through the atmosphere, their paths are deflected in accordance with the laws of optics formulated by Descartes in the early 17th century. The net result is that the apparent position of a star in the sky is not its actual position. This disparity is due to atmospheric refraction, which makes the star appear slightly higher in the sky than it actually is. Refraction is greatest near the horizon since the light must traverse the longest stretch of atmosphere, and diminishes as the star rises higher in the sky. A further consideration is that the degree of refraction is tied to wavelength, so that the positional error also varies depending on what wavelength is being

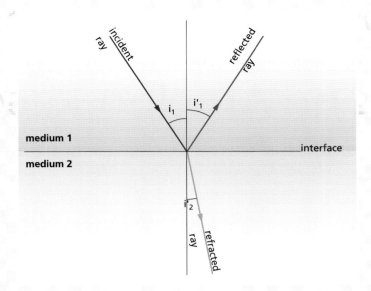

● Diagram showing how a ray of light, passing from a transparent medium of refractive index 1 to a medium of refractive index 2, can be either reflected or refracted depending on its angle of incident. Descartes' law makes it possible to calculate the angle of reflection and refraction relative to the angle of incident.

observed (the chromatic effect). Fortunately many of these effects can be minimized through correctors in the optical path of the telescope or the support instrumentation.

ATMOSPHERIC TURBULENCE

The Earth's atmosphere is not a homogeneous medium, but consists of numerous convection cells ranging in size from a few millimeters to several hundred meters. These cells are altered by changes in temperature and wind currents, and they in turn affect any electromagnetic radiation passing through the air. When starlight enters the atmosphere it encounters numerous cells and is deflected by them many times. This causes distortion in the wave front of light rays, or the surface perpendicular to their angle of entry, and is referred to as turbulence (or seeing) by astronomers. This effect is particularly noticeable in the visible and near-infrared ranges, resulting in telescopic images that seem to "boil" or "twinkle." Images appear far more spread out as a result than they do when observed from space, making it more difficult to distinguish really fine detail.

For many decades atmospheric seeing was considered the main factor limiting telescope performance in the visible and near-infrared range. In the 1990s, however, astronomers started using adaptive optics to compensate for this and all optical and near-infrared telescopes were equipped with this new technology. Seeing is also a prime factor in site selection for new observatories, and Mauna Kea in Hawaii, and Cerro Paranal in Chile, are locations with exceptionally steady atmospheric conditions. ■

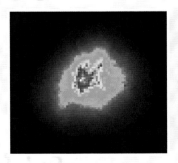
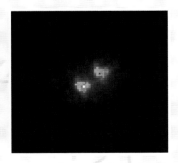

● Atmospheric turbulence or "seeing" degrades or blurs telescopic images of astronomical objects. Adaptive optics compensate for these atmospheric effects and restore image quality, as shown here by a double star imaged with the Adonis telescope at ESO: on the left without correction and on the right with correction.

The Arecibo Radio Telescope

The scene is one of breathtaking beauty and eerie strangeness. Orion has been chasing the Pleiades across the mid-winter sky for hours, yet the weather is balmy. As three gigantic totem poles rise skyward in this serene American setting, they appear tilted, yet completely rigid, glowing in the dim red light. Although bright moonlight illuminates the scene, it does not penetrate the dark forest below, which is filled with a cacophony of piercing cries and ill-defined rustlings. A steady metallic rumble can be heard, however, emitted by a small capsule suspended from an immense steel spiderweb high above ground. This is the steady beat of the compressor pump, which keeps the electronic receivers almost as cold as the interstellar space whose distant radio signals they are collecting.

Arecibo is probably the most exotic astronomical observing facility in the world. After all, what could be more astounding than astronomers trying to detect signals from other inhabited worlds? Evoking images from Clarke, Asimov, Lovecraft and Bradbury, researchers here hope and actually believe that they might someday make contact with alien civilizations on other planets.

Arecibo's giant collecting dish is by any measure the largest in the world. The 300 m diameter radio telescope was built on the island of Puerto Rico in 1963 for the National Science Foundation by Cornell University. Although specializing in the study of pulsars and quasars, the telescope has also been used as an upper atmospheric probe and to bounce radar signals off other planets. The size of a small lunar crater, the huge collecting dish was built into a natural, nearly circular mountain depression just a few dozen kilometers from the Caribbean.

Even today, this is an extraordinarily powerful telescope; ten times more sensitive than its 100 m counterparts at Effelsberg and Green Bank, it has secured a rich harvest of observations and

UNITED
STATES

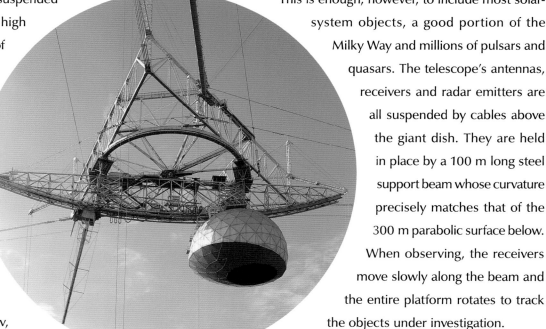

■ *The giant radio telescope at Arecibo has observed the Universe for more than four decades. In 1996 it was equipped with a new secondary antenna support system and a much more sensitive receiver at its focal point (shown here at lower right).*

discoveries in most areas of astrophysics. Since this instrument is far too large to be completely mobile, it was built with a very innovative design. The collecting dish itself is fixed and always pointed toward the zenith. However, an array of receptors and antennas suspended above it can be tilted and moved to track objects for extended periods. This semimobile arrangement has clear limitations. The telescope is limited to covering a section of sky near the 40° latitude of Puerto Rico and about +18° of declination. This is enough, however, to include most solar-system objects, a good portion of the Milky Way and millions of pulsars and quasars. The telescope's antennas, receivers and radar emitters are all suspended by cables above the giant dish. They are held in place by a 100 m long steel support beam whose curvature precisely matches that of the 300 m parabolic surface below. When observing, the receivers move slowly along the beam and the entire platform rotates to track the objects under investigation.

Arecibo's impressive dimensions are hard to visualize. The collecting dish is about 50 m deep and, with a circumference of about 1 km, provides a great jogging trail for the more athletically inclined astronomers. Its 40,000 aluminum surface plates cover an area of 7 hectares and are supported to an accuracy of 3 mm by a thin mesh of cables. This allows observation at wavelengths ranging from .06 m to 6 m. The dish also serves as an effective greenhouse for a lush garden of ferns, begonias and orchids below, as well as a refuge for thousands of insects, birds, bats, giant toads and all sorts of reptiles. A kilometer-long catwalk connects the three towers that support the receiver platform 130 m above the telescope's metal collecting surface. Visitors often fear negotiating that catwalk, not only due to vertigo but also because it sways in the wind without warning, just after the alarm has sounded.

■ *The Arecibo radio telescope as seen in moonlight. A wall of fog envelops the great metal dish, some 300 m in diameter and 50 m deep. The radio telescope itself is fixed but the receivers and antennas on the platform above it can be adjusted to follow the apparent motion of celestial objects in the sky.*

ARECIBO'S NOBEL PRIZE

Well before joining the ranks of classical radio telescopes to investigate radiation from nebulae, pulsars and quasars, as well as searching for signals from potential extraterrestrial civilizations, the great Arecibo dish broke new ground as a powerful radar emitter, capable of bouncing signals off the Moon and the planets. This most powerful radar telescope in the world achieved many firsts. These included the first accurate determinations of Venus and Mercury's rotation periods, the first surface relief maps of the Moon and Mars (before the Apollo and Viking landers) and the first evidence of cratering on Venus, including the 11,000 m Maxwell volcano. Arecibo also obtained the first radar images of comets and asteroids, revealing the strange peanut and dumbbell-like shapes of the asteroids Castalia and Cleopatra.

■ Weighing 900 metric tons, the receiver platform is suspended at the focal point of the Arecibo dish by several kilometers of cables. Some 26 separate motors ensure that the 100 m long support beam and receivers can turn slowly in a vertical plane.

Arecibo scientists have also distinguished themselves in the realm of regular radio astronomy and have taken full advantage of the telescope's 300 m aperture and exceptional resolving power to study pulsars. This led the American astronomers to discover that the rotation of the Crab Nebula pulsar is progressively slowing down, a few billionths of a second per second, and that the pulsar PSR 1397+214, an object less than 10 km in diameter, has the shortest period of rotation observed to date, an astonishing 642 times per second.

Arecibo has also been pivotal in studies of one of the most important objects in modern cosmology, the famed double pulsar PSR 1913+16 in the constellation of Aquila. American astronomers Joseph Taylor and Russell Hulse have investigated this object from Arecibo and shown that the progressive slowdown in

pulsars; in the latter, it can send a megawatt radar pulse to the surfaces of asteroids and planets.

the orbital period of this binary pulsar is due to their gradual inward spiraling. This slowdown was attributed to the dissipation of great amounts of energy as gravitational radiation, a phenomenon predicted by the theory of general relativity. This discovery earned the two researchers the Nobel Prize for physics in 1993.

Arecibo is not known primarily for this fundamental discovery, however. What is more widely remembered is that it was at this facility, some 30 years ago, that humanity's first attempt to communicate with extraterrestrial civilizations was initiated. On November 16, 1974, this radio telescope sent a brief signal toward M13 in Hercules, a globular cluster containing several hundreds of thousands of stars about 25,000 light-years from Earth. This two-minute, 49-second signal, coming from an obscure point in the Universe,

■ *A convoluted 1 km long route passes beneath the 7 hectare surface of the telescope's collecting dish. A fine metal mesh supports the 40,000 reflective panels to an accuracy of 3 mm.*

the Earth, near an unremarkable star, the Sun, would nevertheless appear some ten billion times brighter than any star, at least at a wavelength of 12 cm! This powerful 450,000 watt radio signal passed the orbit of the Moon in about one second, the orbit of Mars five minutes later, and thirty-five minutes later the orbit of Jupiter. Five hours after that, the message passed the planet Pluto and left the Solar System forever. It carried with it a "postcard" from planet Earth encapsulating the history of its inhabitants and their hopes that they are not alone in the Universe.

As the Arecibo message streams toward the Hercules cluster at the speed of light, having traversed some 30 of the 25,000 light-years by 2004, teams of researchers working under the auspices of the SETI Institute are actively using this radio telescope to seek and decipher potential signals from advanced alien civilizations, unsuccessfully … so far. ■

■ The radio telescope at Green Bank is 150 m high and has a total movable mass of 7,300 metric tons. The very impressive altazimuth mount supports a dish more than 100 m in diameter, providing astronomers a total collecting area of more than 8,000 square meters.

The Green Bank Radio Observatory

On a dark and rainy night in 1988 in the wooded hills of West Virginia, the giant radio telescope at Green Bank was observing a distant pulsar, oblivious to the sheets of pouring rain. This was the largest radio telescope in the world at the time, with a dish 91 m in diameter. Then, quite suddenly and without warning, the entire structure collapsed. The next morning, incredulous technicians and astronomers faced a pile of several thousand metric tons of twisted metal. Fortunately no one was injured in this accident, since the instrument was remote-controlled from a building several dozen meters away. Only later did it become clear that this magnificent white telescope, built in 1970, had simply worn out, weakened after thousands of rotations and alignments in the course of its work. One of its main truss supports caved in that night and brought the entire structure down with it. The staff of the U.S. National Radio Astronomy Observatory, stunned by the accident, decided immediately to replace this telescope as quickly as possible with an even larger instrument.

The lush meadows of Green Bank have been home to a variety of radio telescopes for nearly half a century. The first one, installed in 1958, was equatorially mounted. It weighs about 200 metric tons and is 26 m in diameter. This instrument gained fame as the first radio telescope used by Frank Drake in 1960 to search for signals from extraterrestrial civilizations. A few years later, a pair of identical radio telescopes was installed at the observatory and linked via a 2,400 m baseline to form the Green Bank interferometer. In addition to its astronomical investigations, this interferometer also served as a test bed for the future VLA (Very Large Array). In 1965, a 42 m, equatorially mounted dish was installed. Weighing some 2,500 metric tons, this massive instrument was intended

UNITED STATES

■ *The Green Bank radio telescope is the largest fully steerable dish in the world. In 2002, this instrument discovered the pulsar PSR J0205+6449, the vestige of a supernova explosion in Cassiopeia that was observed in 1181.*

for work in the 2 to 40 mm wavelength range. In 1970 the great 91 m telescope was placed in service…and the rest is history.

In 2001, less than six years from design to completion, the very advanced Green Bank Telescope (GBT) was installed to replace the destroyed telescope. Funded by the National Science Foundation, this is the largest fully steerable radio telescope in the world. The GBT weighs more than 7,300 metric tons and is almost 150 m high! Equipped with a 100 m parabolic dish, it is altazimuth-mounted and can observe at wavelengths from 3 mm to 3 m. The design of this instrument is really quite unusual. Unlike more traditional radio telescopes, the GBT main dish is asymmetric in shape, measuring 100 x 110 m, and its secondary support arm is located to one side, allowing it to focus the radio beam directly onto the receiver. This configuration offers many advantages. First, the entire 8,000-square-meter collecting area of the primary dish can be utilized. Second, there are no secondary metal obstructions to degrade or weaken incoming radio signals. Lastly, despite its size and weight, the GBT was built to maintain an unprecedented surface accuracy. The collecting dish consists of 2,004 separate panels mounted on small servomotors. A system of computer-controlled lasers monitors the shape of the dish at all times, and automatically adjusts each panel six times per second to maintain its overall surface accuracy to less than .5 mm.

The GBT promises to be an instrument of exceptional precision and sensitivity. It will be used primarily for studies of some of the weakest radio sources in the Universe. This will be greatly facilitated by its location, which is particularly well shielded from extraneous radio noise; the county of Pocahontas, with an area of 20,000 square kilometers, has been designated a zone of "radio silence." ■

The Very Large Array

It is dawn on the Plains of Saint Augustin in New Mexico. The last quarter Moon is fading in the morning sky and the dry desert vegetation is coated with a thin layer of frost. A coyote finally retires to his burrow after hunting and howling half the night. Although its latitude is in line with the Sahara Desert, the temperature here hovers around –10°C. The Plains of Saint Augustin are a curious geologic feature: at an elevation of 2,200 m they are perfectly flat, ringed by mountains higher than 3,000 m. This high New Mexico desert, midway between Albuquerque and Las Cruces, is far from civilization. Once an ancient lakebed, it is now covered with yellow vegetation and visited only occasionally by cowboys and cattle. There, under crystal clear skies and sheltered from most electromagnetic radiation, we find one of the most powerful astronomical instruments in the world.

The Sun has still not warmed the countryside, but its first rays cast a golden glow on the metal framework of an immobile radio telescope pointing silently skyward. An identical

■ One of the world's largest interferometers is located at an elevation of 2,200 m on the Plains of Saint Augustin. The VLA is protected from extraneous radio interference by the ring of high mountains that surround it.

structure appears a little farther away, then a third and another one farther still. All of them are linked underground by cables and on the surface by a pair of railway tracks extending to the horizon. Twenty-seven radio dishes, extended over an area of nearly 1,000 square kilometers, are all pointed rigidly at a single point in the sky. A coyote on one of the rail ties doesn't bother to move. Apart from these strange metal structures, the plain appears empty. Highway 60, winding through the array and linking the towns of Magdalena and Datil, is deserted.

The Very Large Array (VLA) is in the middle of an observing session; its 27 dishes are focused on the plasma jet emitted by a giant galaxy some 50 million light-years from us. In the control center the operator is surveying the scene with binoculars, and notes that an eagle is still nesting on the scaffold of the eighth radio telescope. The fact that the position of its nest is regularly changed does not seem to bother the bird at all. A single individual can operate this entire complex array.

When the decision was made by the National Radio Astronomy Observatory (NRAO) in 1961 to build the VLA, the main goal was to generate visible images using radio wavelength data, something not possible with smaller radio telescopes at the time. Since the aperture of any telescope is dictated by the wavelength at which it observes, the longer the wavelength the larger the aperture required. At a wavelength of 7 mm, for example, a single 25 m dish like those used at the VLA has an approximate resolving power of only 60 arcsec. That's no better than the human eye in the visible range. Things get even worse at longer wavelengths. The resolving power of a 25 m dish observing at .21 m is so poor that objects like the Sun, Moon, star clusters and nebulae can barely be distinguished from background noise. In short, to attain resolution at radio wavelengths comparable to that obtained in visible light would theoretically require radio telescopes from a thousand to a million times larger than their optical counterparts. Since this is clearly impossible, radio astronomers have turned to interferometry and created synthetic instruments with much larger effective apertures. The designers of the VLA have pushed this principle to near perfection, and the VLA can be described as the largest and most unusual imaging instrument in the world.

■ *Occasionally the VLA is used to track NASA space probes, or to receive radar signals sent by the Arecibo radio telescope to asteroids and planets.*

A GIGANTIC IMAGING DEVICE

After several years of planning, the ambitious VLA project received U.S. congressional approval in 1972. Work began the following year. The full array was completed in 1980 and since then has produced numerous astronomical images in the 7 and 4 mm wavelength ranges. The array of 27 individual 25 m dishes can be moved along rails in a huge Y-shaped layout. Every three months on average, the 235-metric-ton dishes are moved into a new configuration by a powerful transporter engine.

The VLA can be structured into four different Y-configurations, simulating apertures of 1, 3.6, 10 and 36 km respectively, and providing correspondingly higher levels of resolution. All dishes work in unison. Readings are obtained from one pair of telescopes at a time, and combined as interference patterns. After readings have been obtained from all 361 pairs of dishes, the data are used to produce computer-generated images of the objects studied. The sensitivity and resolving power of any interferometer is largely a function of the number of individual collectors in the array, either radio or optical. Because the VLA is composed of so many elements, it functions almost as well as if it were a regular large-aperture radio telescope.

The VLA is least "diluted" in its most compact configuration, where it functions like a 1 km aperture instrument with a total

■ Placed into service in 1980, the Very Large Array (VLA) has undergone constant improvement and upgrading by the technicians and engineers of the U.S. National Radio

collecting surface equivalent to a single 130 m dish. Maximum resolution, however, is attained when the VLA is in its most extended configuration: 36 km. In that mode it can attain resolution in the order of 0.05 arcsec when observing at .01 m (comparable to the Hubble Space Telescope in the visible range) and 1 arcsec at a wavelength of 21 cm. The VLA can usefully observe at wavelengths extending from 7 mm to 4 m.

When the VLA is in its widest configuration, its component dishes are extended in all directions as far as the eye can see. In the midday heat the furthest telescopes shimmer like distant phantoms. They present an awesome spectacle when, quite suddenly, they reorient in unison toward a different point in the sky. Seeing these giant structures moving in synchrony as if they were alive is truly a magical moment.

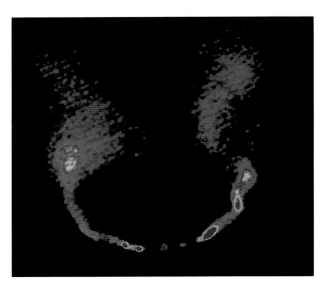

■ This shows a VLA image of the galaxy NGC 1265 in the constellation of Perseus. The hundreds of billions of individual stars in this galaxy are not evident in this image; however, the compact galactic nucleus is visible at the bottom center with two immense plasma jets emanating from it.

In its 25 years of operation, the VLA has been used by thousands of investigators and carried out over 10,000 different observational programs. Thanks to its wide field, sensitivity and high resolving power, the array has generated some truly unique images of the center of our galaxy, and has provided convincing evidence of a massive black hole hidden in the Milky Way's core. The VLA has likewise provided much information on unusual binary stars, neutron stars and black holes in other galaxies. In recent years it has been involved in research on protoplanetary systems in other stars.

For a number of years, the U.S. National Radio Astronomy Observatory planned to upgrade all VLA dishes and link them by optical fibers instead of cables. Above all, astronomers want to enhance both the sensitivity and

Astronomy Observatory. The VLA is shown above in its most compact configuration, with all 27 dishes in a Y-shaped pattern measuring 1 km in diameter.

resolving power of the array. In 2000 the National Science Foundation allocated nearly $150 million toward this goal.

This project known as EVLA (Expanded Very Large Array) will take place in two stages. First, the number of dishes will be increased, and second, the VLA will be linked with the other radio telescopes of the VLBA. In 2001 the VLA was linked by optical fibers to the 25 m dish at Pie Town, about 50 km away. Another VLBA element near Los Alamos, New Mexico, will also join the array, as well as eight new radio telescopes scattered across the United States. By 2010, the EVLA will encompass some 40 dishes in total, forming a 300 km interferometric array with a collective surface area equivalent to a single 160 m dish. This enormous array will attain a maximum angular resolution of 0.005 arcsec.

■ *The galaxy 3C 75 is seen here as imaged by the VLA at .2 m wavelength. This widefield image resolves detail down to 4 arcsec, and was obtained by combining data from three of the four possible interferometric configurations. A massive black hole is located at the center of this galaxy, close to the point of origin of the plasma jets.*

When completed, this will be the largest and most powerful radio interferometer in the world. Capable of resolving detail down to 25 million km at a distance of 100 light-years, it will be used principally for research on protostar formation in dark nebulae and protoplanetary disks around nearby stars. It is anticipated that the expanded "super-VLA," with its wide field of view and very high resolution, should resolve detail in the order of 6 light-hours at the Milky Way's core, 28,000 light-years away. Finally, because of its exceptional sensitivity, the expanded VLA will be able to study galaxies with redshifts of $z = 5$ to 10. This goes back to when the Universe was only a few hundred million years old. Particular attention will be paid to when and how the first massive black holes were formed. ■

> Spectroscopy: the detailed analysis of light

*O*f the many techniques developed to analyze light, spectroscopy has provided the most insight and information. Since its first application over two centuries ago, the science of spectroscopy has led to fuller understanding of atoms and matter, as well as to the discovery of entirely new elements. In astronomy, spectroscopy has not only helped determine the chemical composition of celestial objects with great precision, but has also provided information from afar on their distance, velocity and temperature. Spectroscopy was also key to the discovery of the recession of the galaxies and the detection of the first extrasolar planets. Spectroscopy continues to be an indispensable tool in astrophysics.

THE PRINCIPLES OF SPECTROSCOPY

When a beam of white light passes through a glass prism it is dispersed into its component colors or spectrum. Prisms were the first spectroscopic instruments in history and are still used extensively today. Even the most modern spectroscopes

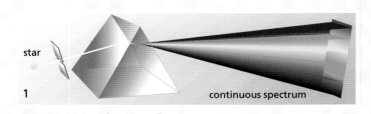

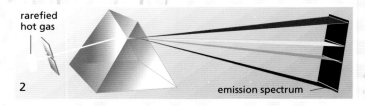

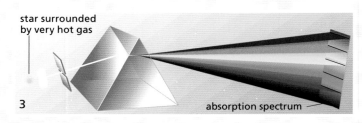

● A prism bends and disperses white starlight into its component colors, much like a rainbow. This is known as a continuous spectrum (1). When a gas is heated to high temperature, it emits very bright spectral lines (2). When gas is located in front of a stellar light source, it absorbs portions of the stars' spectrum and these appear as dark lines (3).

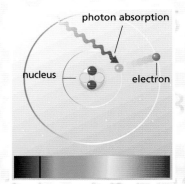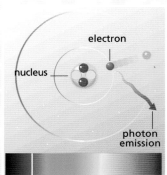

● When the electrons in an atom jump from a lower energy level to a higher one they absorb that energy. This appears as a dark band in the spectrum and is called an absorption line. The opposite occurs when electrons jump from a higher energy level to a lower one and energy is released. This appears as a bright emission band in the spectrum.

still operate on the basis of bending and dispersing radiation as a function of wavelength.

At visible wavelengths, the Sun's spectrum exhibits numerous dark absorption lines superimposed over its continuous spectrum. These lines are due to various chemical elements in the solar atmosphere that absorb photons of specific wavelengths. In order to better understand this we must look into the structure of atoms. All atoms consist of a core or nucleus of protons and neutrons surrounded by whirls of orbiting electrons in shells of different energy levels. Each electron occupies a well-defined orbit and energy level. Atoms become excited when struck by photons, for example, and the absorption of that energy can

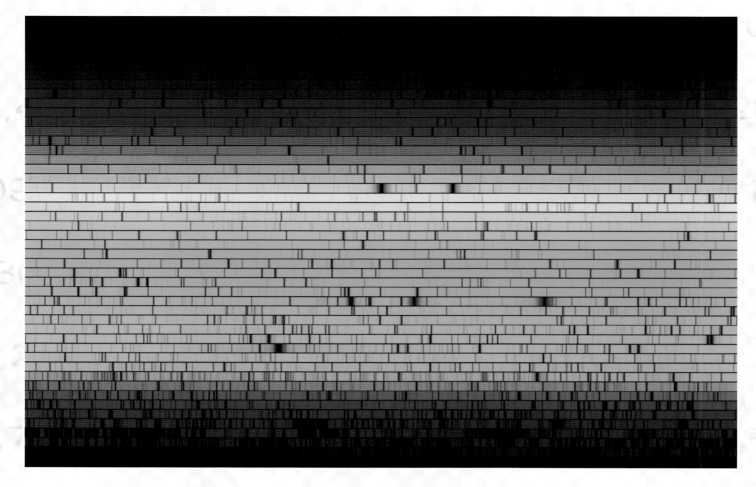

● This high-resolution spectrum of the Sun was obtained with the most powerful solar spectrograph in the world, installed on the McMath-Pierce telescope at Kitt Peak Observatory in Arizona. Each of the 50 spectral segments shown covers only 6 nm, but collectively include the full 400 to 700 nm wavelength range of the visible spectrum. More than a thousand absorption lines are shown, indicating the presence of hydrogen, helium, lithium, beryllium, boron, carbon, nitrogen, oxygen, fluorine, neon, sodium, magnesium, etc., in the solar atmosphere.

boost electrons with the lowest binding energy to a higher energy level. The wavelength of the incoming photons determines the energy level to which electrons are sent to or return from. When electrons jump to a higher energy level, this is manifested as spectral absorption lines. Excited electrons, however, do not stay at such high-energy levels very long, but eventually return to their ground state. In so doing they emit their excess energy as photons of the same wavelength they absorbed. This is manifested as emission lines in the spectrum.

According to the laws of quantum mechanics, only a limited number of energy levels is possible for any given element. For example, when a ray of white light passes through a gas cloud, hydrogen atoms absorb only photons whose wavelengths correspond to its permissible energy levels. The light that emerges from such a cloud lacks photons of those wavelengths, resulting in spectral absorption lines characteristic of hydrogen. In a similar manner, light passing through an oxygen-rich medium will exhibit spectral absorption lines characteristic of oxygen, and so on for various other elements. That is why each chemical element has its own characteristic spectral signature.

Most of the absorption resulting from electron transitions in atoms, ions and molecules appears in the visible,

ultraviolet and infrared ranges of the electromagnetic spectrum. However, with some extremely high-energy transitions, such as highly ionized iron, absorption lines appear at X-ray wavelengths.

When electromagnetic radiation interacts with larger molecules and interstellar dust grains, similar types of quantifiable energy exchanges take place, which also produce spectral absorption and emission phenomena. For example, the rotating and vibrating energies produced by molecules in space can be detected at infrared and submillimeter wavelengths. Likewise, high-energy nuclear transitions and nuclear fusions can be detected as gamma and X-rays respectively.

ASTROPHYSICAL APPLICATIONS OF SPECTROSCOPY

Spectroscopy has helped us identify the many different atoms, ions, molecules and dust components present in celestial objects and the interstellar environment. Beyond mere identification, however, spectroscopy has also provided important quantitative data on the elements and compounds detected. For instance, detailed information on elemental abundances, temperatures and other key physical characteristics can be

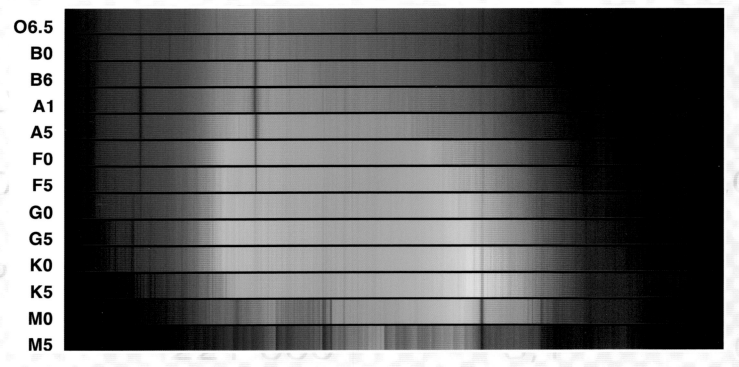

O6.5
B0
B6
A1
A5
F0
F5
G0
G5
K0
K5
M0
M5

● This figure shows the main stellar spectral classes, ranging from hot, young blue stars (types O and B) at top, down to dwarfs and cool red giants (types K and M), with solar-type stars (G) in the middle. All spectra were taken with the .9 m telescope at Kitt Peak. The spectra of G0 and G5 stars can be compared to that of the Sun (page 71) taken with the McMath-Pierce telescope.

obtained from the intensity and widths of their respective absorption lines. This has led to our fuller understanding of the chemical composition and temperature of the atmospheres of the Sun, other stars and planets, the interstellar medium, other galaxies, and so forth.

Another very important application of spectroscopy is the use of the Doppler effect to measure the speed and motion of various celestial objects. This has provided crucial information on the dynamics of planets, stars, galaxies and quasars. Edwin Hubble used these approaches in 1929 to show that galaxies are receding and that the Universe is expanding. Spectroscopy also led to the discovery of the first exoplanet in 1995.

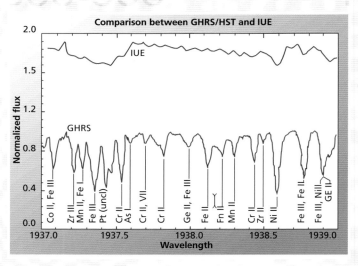

● A short portion of the ultraviolet spectrum of the star Xi Lupi, as shown, taken at medium resolution (12,000, upper trace) and high resolution (87,000, lower trace). Only the latter resolves the many absorption bands characteristic of various ions in this star's atmosphere.

THE POWER OF SPECTRAL RESOLUTION

The resolving power of any spectrograph is determined by its ability to separate electromagnetic radiation into segments ($\square\lambda$) that are small relative to their wavelengths (λ). Resolving power is expressed as $\lambda/\square\lambda$. The higher this value, the greater the ability of the spectrograph to disperse different wavelengths and resolve closely spaced spectral lines. Spectrographs using glass prisms can typically separate several dozen lines in the visible range and several hundred thousand bands when fine glass gratings are used. For visible light, the UHRF spectrograph at the 4 m Anglo-Australian Telescope (AAT) holds the record, with a spectral resolution of one million.

In astronomy the type of spectrograph used depends on the information sought, the type of object investigated and the sensitivity levels required. Let's take two specific examples. Spectral resolution of a few thousand is sufficient to measure the speed of recession of galaxies or quasars, while resolution in the order of 50,000 is needed for detailed assessment of the chemical composition of stars or the interstellar medium. Also, since a high degree of dispersion is required for high-resolution spectroscopy, only relatively bright objects are suitable for this type of analysis. As a result, the composition and temperature of nearby stars and nebulae is better characterized than those of similar objects at greater distances.

Naturally, as telescopes have increased in size, higher-resolution spectroscopy has been applied to increasingly fainter

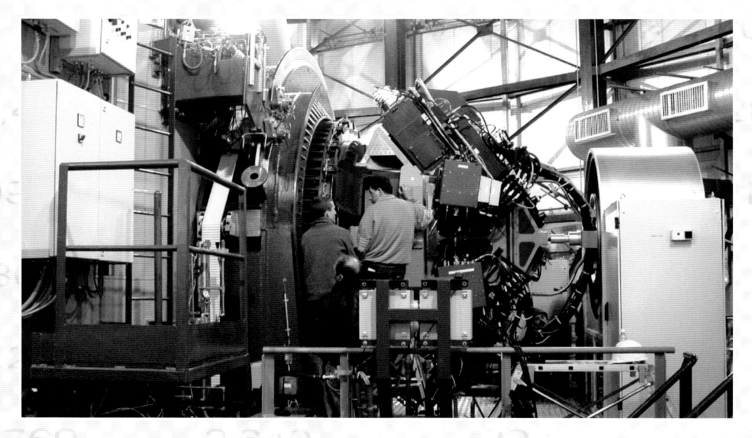

● In 2002, the new spectrograph-imager Vimos was installed on the 8.2 m Melipal telescope of the European VLT array. This instrument is capable of recording the spectra of nearly 1,000 galaxies to 23rd magnitude simultaneously, with exposures of only 15 minutes. Vimos is mounted at one of the telescope's Nasmyth foci and slowly rotates with it during the long exposure times needed to register spectra.

objects. It is now possible to obtain at least medium resolution spectra with the new 10 m class telescopes of such dim objects as brown dwarfs and quasars, something that was simply not possible with smaller instruments.

WIDEFIELD SPECTROSCOPY

For the most part, traditional spectroscopy only recorded spectra of one object at a time. This meant taking separate exposures of stars in an open cluster, or of all members of a group of galaxies, and required several observing sessions

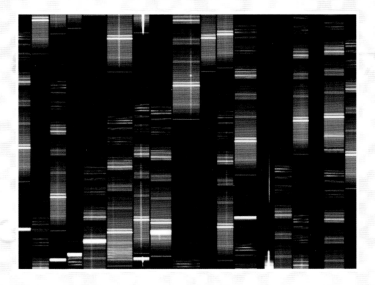

● The spectrograph-imager Vimos will survey more than 150,000 galaxies to construct a three-dimensional map of the Universe.

and a prohibitive amount of telescope time. All this changed, however, with the development of multi-object spectrographs, which can obtain low to medium resolution spectra of tens or even hundreds of objects at once. One technique uses bundles of optical fibers positioned over several preselected objects within the field of view. The 2dF spectrograph at the Anglo-Australian Telescope (AAT) used this method to pre-position 400 optical fibers and obtain concurrent spectra of 400 galaxies. A similar method has been used since 2002 with the OzPoz spectrograph at the VLT with some 560 optical fibers. This approach is particularly well suited for spectroscopic surveys of star and galaxy clusters.

Spectroscopy of diffuse objects or regions lacking well-defined borders is particularly difficult. Several approaches have been tried to overcome this. A Fabry-Pérot imaging spectrometer can study properties of extended sources such as groups of gravitationally microlensed objects. The instrument records spectra of the individual objects and then integrates the information across the entire field of view. The OASIS spectrograph with the Canada–France–Hawaii Telescope (CFHT) on Mauna Kea was built along these lines, as was the Vimos spectrograph for the VLT. Instruments like these extend our reach well beyond the capabilities of traditional spectroscopy and are likely to be used extensively in future, especially for extragalactic astronomy. ■

The Very Long Baseline Array

A mid the glaciers and ocher-gray volcanic mounds of Mauna Kea, total silence is the most impressive sensation. The winds that buffet the mountain summit some 500 m higher rarely stir this seemingly lifeless geologic expanse. And yet on this fine April afternoon something is moving amid this otherwise static scene. A sharp metallic groan can be heard in the cold dry air, sounding like some mythical creature, which dies away as quickly as it came. A closer look reveals a gleaming-white 30 m parabolic dish slowly turning skyward. It is a huge radio telescope, firmly anchored to the ground and clearly under remote control. Though clearly inanimate, the instrument does seem strangely alive with clicks and creaks as it seeks some distant target in the sky.

This solitary Mauna Kea antenna is not an independent radio telescope but just one of several elements of a much larger complex known as the Very Long Baseline Array (VLBA). This American facility is not only the largest radio interferometer in the world, but also the largest astronomical observatory on the planet. The VLBA consists of 10 identical 25 m diameter dishes, altazimuth-mounted and covering all portions of the northern sky.

The VLBA covers the full geographic extent of the United States, measuring 8,000 km east to west, and 4,000 km north to south. Since the elements of the array are so widely distributed, astronomers across the country have access to this unique facility. The easternmost point of the array is on a sandy beach in the Virgin Islands in the Caribbean, and the westernmost element is located atop Mauna Kea in Hawaii. The remaining elements of the enormous interferometer cover much of the North American continent, extending from New Hampshire to Texas, Iowa, New

■ *The heart of the 25 m Mauna Kea radio telescope. All 10 elements of the VLBA are equipped with identical receivers and recorders. Since all data are recorded on magnetic tape, the array can observe for 12 consecutive hours at a time.*

Mexico, California and Arizona. The VLBA observes at wavelengths between 7 mm and .9 m, a region of the radio spectrum where stars are basically "invisible," but where areas of nebulosity heated by supergiants radiate strongly. However, this powerful instrument will also examine the more exotic objects in the Universe, namely pulsars, quasars and black holes.

As we saw before, our ability to resolve fine detail in astronomical images depends both on the effective aperture of a radio telescope or mirror, and the wavelength at which we observe. The longer the wavelength the larger the aperture required. At visible wavelengths between 0.5 and 1 μm, where our eyes are most sensitive, both the Hubble Space Telescopes and the largest Earth-based instruments can resolve detail to about 0.05 arcsec. Depending on what wavelength it is observing at, the resolving power of the VLBA is an incredible 0.03 to 0.0005 arcsec, or about a hundred times better than the HST. With visual acuity like this, someone in Paris, France, could read this very book from as far away as Dakar, Senegal, or someone in San Francisco could read this book from as far away as New York City. Put into a more astronomical context, 0.0005 arcsec equates to detail of less than 1 m on the Moon as seen from Earth!

As with all radio interferometers, each VLBA dish collects separate signals from one target object. These signals are then combined into a single image formed by a "virtual" telescope the size of the entire array, in this case 8,000 km. That explains its extraordinary resolving power. On the other hand, the VLBA does not have the sensitivity that a true 8,000 km radio telescope would have, since the combined collecting area of its 10 dishes is the equivalent of a much smaller 80 m telescope. This is roughly

comparable to those at Parkes (64 m), Jodrell Bank (76 m), and Effelsberg and Green Bank (both 100 m in diameter). Collecting areas of that size are sufficient, nonetheless, to capture very weak signals from distant galaxies in just a few hours of observing time.

FROM THE VIRGIN ISLANDS TO HAWAII

The VLBA nerve center is located in Socorro, a small town in central New Mexico. From there operators work the giant radio telescope around the clock while monitoring the status of each separate dish and the weather conditions at each site. Weather can vary significantly at each location. In the fall of 2000, as the Virgin Islands dish was baking in the midday sun at its beach location under calm blue skies, the Mauna Kea telescope at the other

■ Elements of the VLBA extend from the Pacific to the Caribbean across all regions of the United States. From the edge of the Atlantic to the top of Mauna Kea, Hawaii, all 10 radio telescopes observe astronomical objects in unison, around the clock.

end of the array went into automatic shutdown mode. This was to avoid damage from the fierce tropical storm that had struck Mauna Kea that morning. A few hours later, the delicate surface of its parabolic dish was completely coated by several metric tons of ice.

When everything functions properly, as it usually does, the operators direct all 10 VLBA dishes at their target in the sky. Although all observing is carried on remotely, teams of site technicians attend each radio telescope daily. Their job is to maintain all instrumentation, keep the receivers refrigerated and replace magnetic tape on the data-recording computers. Each cassette contains 5,500 m of tape that can store 1.5 megabytes of data per centimeter, or enough for about 12 hours of nonstop operation.

■ *The radio telescope at Pie Town, New Mexico, is used alternately with the VLA and the VLBA. The VLBA can attain a resolution of 0.0005 arcsec when observing at wavelengths between 7 mm and .9 m. The VLBA is often combined with the European VLBI array to increase resolving power even more.*

In addition to insuring that all components of the VLBA point to and accurately track the same object, signals received by each dish must be properly synchronized. This is done by atomic clocks at each site. Accurate to within one second per million years, atomic clocks provide an "absolute" reference time point for the entire array. The data tapes are shipped to Socorro by courier once a week and analyzed by special high-speed decoders. These recover data from each paired set of antennas at the rate of 750 operations per second. Because of the extreme accuracy of the atomic clocks at each site, signals from each receiver are readily differentiated during this analysis.

The VLBA decoding computer first constructs a virtual 8,000 km wide "dish," and then reconstructs images or spectra of the objects observed by the array. To do this correctly, the computers must apply a number of corrections to the signals to simulate a large, cohesive collecting dish. First, corrections must be applied that compensate for differences in the relative elevation of each antenna (which ranges from 0 to 3,700 m), as well as for the curvature and rotation of the Earth. After these adjustments, data from the 45 interferometric pairings must be synchronized to generate the final image or spectrum. It takes several weeks of data reduction before the researchers finally see their results.

OBSERVING A SUPERNOVA EXPLOSION

The VLBA has proven invaluable in clarifying the structure of galactic nuclei. Ordinary telescopes are of limited use in this regard, since the nuclei of very distant galaxies appear as little more than bright points of light at optical wavelengths. The VLBA, however, has been able to penetrate deep into galactic cores and their surrounding regions. Because of this, American astronomers were able to show that the beautiful spiral galaxy NGC 4258 in Ursa Major rotates around a giant black hole. This remarkable object contains the equivalent of about 40 million solar masses packed into a very small volume. It was also discovered that a flattened disk of hot gas rings the nucleus of this galaxy, spinning at about 1,000 km/s. This structure is less than one light-year in diameter, or 0.008 arcsec as seen from Earth, and lies within five light-months of the center of the galaxy. NGC 4258 is about 24 million light-years from us. The VLBA has also resolved detail in the core of M87, a galaxy in Virgo twice as far from us, at 50 million light-years.

M87 has a very small, compact nucleus, with a 6,000 light-year-long plasma jet emerging from it. To observe this in more detail astronomers linked not only the VLBA and the VLA but also the large European interferometer, the VLBI. Observing M87 at a wavelength of 1.3 cm, this combination yielded 0.0001 arcsec resolution, corresponding to just a few light-months, which was an absolute record for that distance. There, too, a rapidly spinning disk of gas was discovered surrounding a huge black hole of several billion solar masses.

The best understood black hole, however, lies at the center of our own Milky Way galaxy. Situated *only* 28,000 light-years from us, the core of our galaxy is not visible at optical

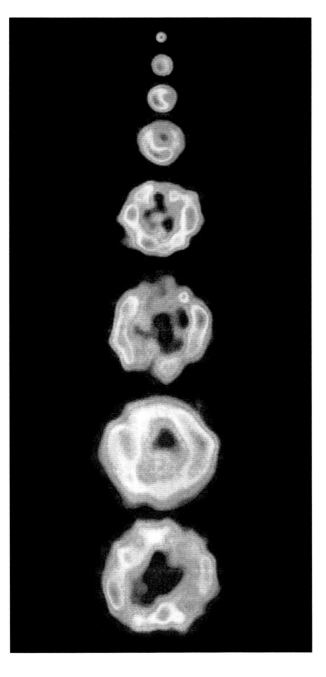

■ *Between May 1993 and November 1997, the joint arrays of the VLA, VLBA and VLBI followed the progress of a supernova explosion in the galaxy M81. The incredible resolving power of this planet-size virtual telescope clearly showed the 10,000 km/s expansion of the supernova's gas shell. These 0.0005 arcsec resolution images were obtained at a wavelength of 6 cm.*

wavelengths because a veritable wall of stars and nebulosity masks it. Officially known as Sgr A*, this region was probed by the VLBA at a wavelength of 7 mm, revealing an elliptically shaped nucleus only 0.004 arcsec across. This corresponds to an actual diameter of less than 500 million km, or about 3.3 times the distance from the Earth to the Sun. This tiny volume hides a gigantic black hole, the equivalent of one million solar masses within the confines of a very strange space/time environment. From time to time, gas clouds and even entire stars venture too close to the event horizon of the black hole, and disappear within it without a trace.

It's the middle of the night in Socorro. The operator on duty is alone and quiet, and nearly lulled to sleep by the steady hum of the Correlator (as the key computer is referred to by those who work here). Perhaps he is dreaming of the surreal landscape that surrounds this giant radio telescope. As the dawn gradually breaks and sunlight illuminates the desert again, the stars vanish against a deep blue sky. The air is cold and a thick morning mist envelops everything. Later that day, a warm shower leaves puddles on the ground, which freeze as the Sun sets again in a cloudless sky.

From time to time the operator glances at the images of quasars emerging at regular intervals from the laser printer. No one before him has ever seen these images from the furthest reaches of the Universe, images that only this powerful Correlator was able to produce. ■

The Mount Wilson Observatory

The telescope glides silently from star to star in the dark of the observatory dome. As its classic equatorial fork mount slowly rotates westward, the telescope's trellis tube outlines several northern constellations. Even as we begin the 21st century, astronomers at Mount Wilson Observatory are still using this renowned instrument, little changed since it was built. Obviously, computers have replaced its original clock drive and electronic receivers have taken the place of eyepieces. However, in the dead of night, time and space might as well have stopped on Mount Wilson. Let's go back to that cold December 8 night in 1908, when George Ellery Hale had finally realized his dream of building the largest telescope in the world, and first pointed the 1.5 m telescope toward the sky. We could even imagine coming across famed astronomer George Richey, who spent four consecutive nights in this dome in 1910 to photograph the galaxy M64 in Coma Berenices. He would have carefully closed his camera at the end of each exposure, and then repeated the process night after night until enough light accumulated to record a good image.

Today as we climb Mount Wilson along a tortuous mountain road, it's easy to forget that some 22 metric tons of telescope parts were originally carried piecemeal by mules along a dusty path all the way to the 1,742 m high summit. Unfortunately, we have also forgotten how beautiful the California skies were in those days. It's difficult to get even a glimpse of the summer Milky Way now, as only the brightest of stars remain visible in a sky flooded by the immense glow of Los Angeles a mere 20 km away.

George Ellery Hale founded Mount Wilson Observatory in 1903. It remains, along with historic institutions like Pic du Midi and others,

UNITED STATES

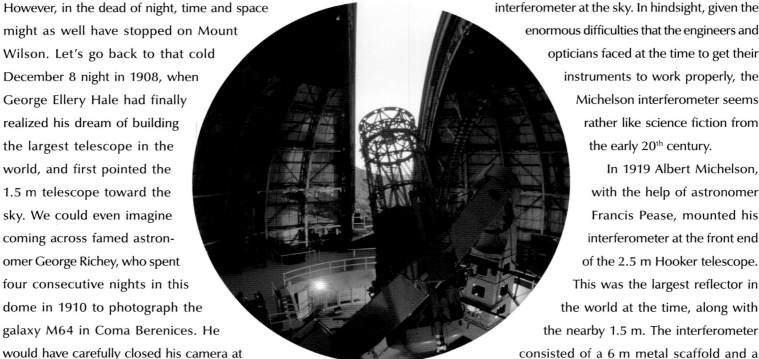

■ *Several fundamental astronomical discoveries were made at Mount Wilson Observatory. The 2.5 m Hooker telescope, largest in the world when installed in 1918, is still in active use today.*

one of the oldest still-active astronomical sites in the world. Despite the extensive light pollution, Mount Wilson astronomers have never given up observing, although today they work mainly on the brightest stars, those least affected by atmospheric transparency. They also continue to specialize in optical interferometry, a long-standing tradition at this observatory. It was here that the American astronomer, Albert Michelson, inspired by the work of the French pioneer Hippolyte Fizeau, first pointed an interferometer at the sky. In hindsight, given the enormous difficulties that the engineers and opticians faced at the time to get their instruments to work properly, the Michelson interferometer seems rather like science fiction from the early 20th century.

In 1919 Albert Michelson, with the help of astronomer Francis Pease, mounted his interferometer at the front end of the 2.5 m Hooker telescope. This was the largest reflector in the world at the time, along with the nearby 1.5 m. The interferometer consisted of a 6 m metal scaffold and a system of mirrors at each end to reflect starlight back onto the telescope's primary mirror. The resulting interference patterns were examined by eye at the telescope's focal plane! Thanks to their pioneering work, however, Michelson and Pease no longer saw stars as mere points of light, but went on to measure their diameters. Quite unexpectedly, these measurements showed that Arcturus and Aldebaran are red giants, and Antares, Betelgeuse, Mira and Ras Algethi are supergiants, some ten to a hundred times bigger than our Sun. These groundbreaking efforts have not been forgotten at Mount Wilson, and over the past 20 years, interferometers have sprouted like mushrooms at the observatory.

In 1982 Mike Shayo rekindled Michelson's torch by building the Mark III interferometer for the U.S. Naval Research Laboratory, and

■ *The dome of the famous Hooker telescope juts out above the California Pines on Mount Wilson. This historical telescope, brought into service a century ago, still operates alongside such modern interferometers as Mark III and CHARA, whose smaller domes are hidden among the trees.*

by 1986 a study of the interference patterns of very bright stars was initiated. The instrument is similar to Michelson's interferometer; with two .25 m mirrors mounted on a north/south axis to bring two images of the same star to a common focus. The distance between the two mirrors can be adjusted from 3 to 30 m to change its resolving power. This interferometer can be used only on stars visible with the naked eye, however, since its light-collecting surface is very small. Among other important findings, it accurately determined the distance of nova Cygnus in 1992, at a distance of 9,200 light-years. It also established the orbital parameters of Alpha Andromedae with remarkable precision. This binary system is located some 96 light-years from us, and the distance between the pair varies from 0.002 to 0.02 arcsec over a three-month cycle.

The ISI (Infrared Spatial Interferometer) was installed on Mount Wilson in 1988 by a team of researchers from the University of California at Berkeley. Working in the infrared at 11 μm, this instrument consists of two movable 2 m mirror flats, which reflect starlight onto two fixed 1.6 m parabolic mirrors. With an adjustable baseline of 4 to

■ *CHARA (Center for High Angular Resolution Astronomy) is one of the most promising optical interferometers. Operating since 2000, CHARA consists of six 1 m telescopes arranged in a circle 350 m in diameter. The domes of the array are shown here before being put in service.*

35 m, the ISI can resolve in the 0.25 to 0.03 arcsec range. In the 1990s it measured the diameter of several stars with unprecedented accuracy, including the supergiants first observed by Michelson and Pease in 1919. The angular diameters of Antares and Betelgeuse are now known to within 0.002, 0.044 and 0.056 arcsec respectively, which translate to about 1 billion km for Betelgeuse and more than 1 billion km for Antares.

ALBERT MICHELSON'S LEGACY

In 2000, CHARA (Center for High Angular Resolution Astronomy) was inaugurated at Mount Wilson. This ambitious $12 million project promises to become one of the most powerful optical interferometers in the coming decade. CHARA is a complex of six 1 m telescopes, arranged in a 350 m wide circle. Designed to observe at both infrared and visible wavelengths, it will resolve detail to 0.0002 arcsec. This will allow it to measure not only the diameters of many bright stars, but also to resolve surface details on such supergiants as Mira, Antares and Betelgeuse — stars first "resolved" on this very site by Albert Michelson more than 85 years ago. ■

■ *Kitt Peak Observatory is one of the largest astronomical centers in the world. The extended summit of this 2,060 m high mountain houses several national instruments, including the 3.8 m Mayall telescope, as well as facilities such as the 3.5 m WIYN telescope, which is operated by several American university consortia.*

The Kitt Peak Observatory

The Sonoran desert of Arizona is populated by forests of giant cacti and dominated by Mount Ioligam. Snow-covered in winter, Ioligam is a sacred mountain and home to the mysterious Itoi, the "ancient brother" of the Papago Indians. For American astronomers though, the long and jagged profile of this mountain is also home to Kitt Peak Observatory, a premier astronomical observing site. National Science Foundation astronomers first set their sights on this nearly ideal location in 1953. However, negotiations between the Papago owners of the land and the so-called "people with large eyes" did not conclude until 1958. Only then was the first stone of this future national observatory set in the Tohono O'odham reservation.

As with many large observatories, Kitt Peak's beginnings were rather modest. Teams of astronomers and technicians first tested the site with a small .4 m telescope to fully evaluate this location in Arizona, 2,060 m high and blessed with 250 clear nights a year. The idea of building a national observatory in the United States goes back to the early 1950s, when the astronomical community came to the realization that most of the large telescopes in America were in the

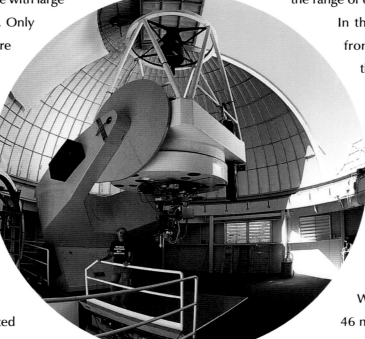

■ The 2.4 m Hiltner telescope is equipped with a spectrograph, a CCD and an infrared camera, all alternately installed at the f/7.5 Cassegrain focus.

hands of private institutions (mainly universities) in California, the richest state. Soon after that, however, domes seemed to mushroom on Kitt Peak. Its first large telescope, the .9 m, was commissioned in 1960 and by 1964 a myriad of telescopes ranging in size from .5 m to 2.1 m had appeared, including a large 2.4 m. All would pale, however, in comparison to the future 4 m Mayall reflector.

This imposing instrument, with a total movable mass of 375 metric tons, was fully operational in 1973 and soon became the model for some half a dozen similar instruments worldwide. Unfortunately, the Mayall telescope mirror was not optically perfect and was subsequently diaphragmed down to an aperture of 3.8 m. Despite this, the telescope remained second largest in the world for many years. The research conducted by the American astronomical

community with this telescope produced many impressive publications, particularly in the realm of extragalactic astronomy and cosmology. By the late 1970s, the Mayall telescope was also among the first to be equipped with a CCD camera.

As with most instruments at that time, this telescope typically attained only 1 arcsec resolution, not an exceptional level of performance for a 3.8 m. Its performance is much better these days, thanks to new CCD cameras and spectrographs, the most powerful of which can attain resolution in the range of 65,000 to 70,000.

In the 1980s, under much pressure from the astronomical community, the National Optical Astronomical Observatory (NOAO) decided to improve matters by building a second instrument of comparable size, a 3.5 m telescope known as the WIYN. This innovative telescope was based on entirely new electro-optical technologies and began operations in 1994. With a moving mass of less than 46 metric tons, it was not only less expensive and more compact than the Mayall telescope, but also performed better.

Kitt Peak Observatory will no doubt occupy an important place in the annals of American astronomy, comparable to La Silla Observatory for Europeans. Although none of the telescopes at this U.S. national facility attained the optical performance levels of the large Hawaiian or Chilean telescopes, Kitt Peak had a huge impact on astronomy. The sheer number of telescopes and quality of instrumentation assured that. Among its most notable discoveries, Kitt Peak provided the first gravitational lensing image of Q0957+561 in 1979. This powerful radio source in Ursa Major, discovered at Jodrell Bank in England, was directly imaged for the first time with the 2.1 m telescope at Kitt Peak. Astronomers showed that its two apparently identical components also had identical spectra. This proved they were actually twin images of a single source, which appeared

double due to an intervening object acting as a gravitational lens. This discovery revolutionized modern astronomy.

THE NEW GALAXIES IN ANDROMEDA

As with many observatories built during the first half of the 20th century, including those at Mount Hamilton, Mount Palomar or even later facilities at Siding Spring or La Silla, Kitt Peak no longer meets the technological and climatic conditions offered to astronomers at superior locations. Because of this, the newest large American instruments, like the twin 8 m Gemini telescopes, are all located in the Chilean Andes and Hawaii. As with many other telescopes of its generation, the Mayall reflector has been converted into a widefield imaging instrument. A large mapping project of the Universe was started in 1997, using a new CCD array called Mosaic. With its 64 million pixels, this

■ *This spectacular WIYN image shows a multitude of stars in the spiral galaxy M101 in the foreground and a very distant galaxy behind it, appearing minuscule in comparison.*

camera covers a field of 36 x 36 arcmin, an area that exceeds the angular size of the full Moon. Operating at the 3.8 m primary 12.4 m focus, this camera records stars and galaxies to 26th magnitude in a one-hour exposure. This produced the National Optical Astronomy Observatory's Deep Wide-Field Survey in 2002, which covers an 18° x 18° field in the constellation of Bootes at six different wavelengths, ranging from visible to infrared. More than five million galaxies to 27th magnitude were recorded in this survey, providing researchers with a unique statistical probe into the cosmos. This Mayall telescope deep-field survey will provide a basis for future probes by the Gemini North telescope of the faintest and most distant objects in the database. Mayall's twin, the 4 m Blanco telescope at Cerro Tololo in Chile, recently began an identical deep-

mirror weighs less than 2 metric tons and features a very short f/1.75 focal ratio. The telescope's total movable mass is less than 46 metric tons.

field survey in a corresponding location in the southern hemisphere.

The Mayall and WIYN telescopes were used jointly in 1998 and 1999, in the discovery of two new members of the Local Group of galaxies. Called And V and And VI, these dwarf galaxies are so faint that they escaped detection prior to this. Astronomers began the search by digitizing photographs of areas around M31, the Andromeda Galaxy, taken with the Mount Palomar Schmidt telescope in the 1980s. The hundred or so "suspect" objects identified this way were then studied in detail by the two large Kitt Peak telescopes. Both V and VI are very small satellite galaxies of M31, located about 900,000 light-years from it and 2.6 million light-years from the Milky Way. These discoveries brought the total number of known galaxies in the Local Group to 43.

■ *Imaged by the WIYN telescope, the magnificent spiral NGC 891 is seen in profile. It is probably very similar to our own Milky Way galaxy.*

Kitt Peak Observatory is not just a nighttime observatory, but has also been home to the most powerful solar telescope in the world since 1961. The housing of the McMath-Pierce telescope resembles a giant sundial. Some 30 m high atop Kitt Peak, the telescope superstructure forms a large triangle whose hypotenuse points directly at the North Celestial Pole. A 2 m diameter plane mirror at the top follows the Sun's path across the sky and diverts its light 20 m below ground to the telescope's 1.6 m primary mirror. With its 86 m focal length, this mirror projects a .76 m diameter image of the Sun, which is then photographed and analyzed spectroscopically. The Fourier transform spectrometer used with the McMath-Pierce telescope holds the world resolution record, with a spectral resolution of 3,000,000! In fact, this telescope is so powerful that astronomers often used it to obtain high-resolution spectra of some of the brighter stars. ■

> Optical and infrared imaging systems

After photos captured by a telescope are filtered, dispersed or sorted, they must be transformed into signals that can be measured and quantified. This requires detectors that can analyze and decipher the data accurately and precisely, using advanced technologies that are continuously modified and improved. No single detector system exists that can meet all needs; a variety of different types of imaging systems are used, tailored to various wavelengths. In the visible and near-infrared range, the photographic plate has been totally replaced by electronic imaging systems, some of which are described below.

SIGNAL DETECTION

The function of any detector is to collect electromagnetic signals and convert them into either images or measurable and quantifiable data. To do this, incoming photons are usually targeted at photosensitive materials that respond to light flux in various ways: for example, through thermal reactions (calorimeters and bolometers); by generating electrons (photomultipliers and CCDs); through photochemical reactions (film and photographic plates), ionization reactions and so forth.

In modern astronomy, the detection and amplification of electromagnetic signals and their accurate conversion into

● In 1980, the very first CCD detectors contained only a few hundred pixels. Modern CCD cameras usually consist of an array of several chips, like the CFH12K camera pictured here, with 100 million pixels.

quantifiable or numerical data are of great importance. In this process, inherent background or instrument "noise" must be filtered out or minimized to provide the strongest and most reliable signal from the object under investigation. Consequently, detectors are generally characterized by such parameters as sensitivity to signal strength, efficiency of signal conversion, signal to noise ratio, thermal stability, size, useful lifespan, etc.

THE BEGINNINGS OF ELECTRONIC IMAGING

The photographic plate was the sole imaging medium available to optical astronomers for a long time and, despite being technically difficult to work with and limited in information content, served their purposes well. Today, however, electronic imaging has not only replaced photography, but has also met the increasingly complex needs of modern astronomy.

Certain materials emit electrons when illuminated by photons of sufficiently high energy. Termed the photoelectric effect, this phenomenon was explained by Einstein at the beginning of the 20th century and formed the basis of the photomultipliers developed in the 1940s and used extensively in astronomy for many decades. There were several advantages to working with electrons in this manner. The number of electrons generated is proportional to the energy levels of the photons striking the receptor, which is information that can be measured quantified, stored and manipulated.

Photomultipliers were designed and built to measure the energy flux across the area observed (termed the photometric aperture). More elaborate systems, such as the Lallemand

● The 2.2 m telescope at La Silla Observatory is equipped with a Wide Field Imager (WFI) camera, containing several CCD receptors in an array of nearly 68 million pixels. The camera is mounted at the telescope's Cassegrain focus behind the primary mirror, whose support cell is pictured here. The WFI camera is specifically designed for widefield imaging.

camera, combined photomultipliers and video capabilities to provide both an image and photometric data simultaneously. Image intensifiers and television were also combined to further enhance astronomical imaging capabilities. In the 1970s, however, all such devices were abandoned in favor of cameras that counted photons directly.

Such instrumentation quickly replaced the traditional photographic plate and eliminated many of the associated drawbacks: extensive setup times, careful chemical development and long and tedious sessions with plate-measuring machines. Above all though, electronic imaging devices were reusable and yielded reproducible results. A new era in astronomical observing was born....

CHARGE-COUPLED DEVICES

Charge-coupled devices or CCDs were developed in 1969 by researchers at Bell Labs in the United States. Made of semi-conductor materials such as silicon, the devices contain chips with numerous microscopic light collectors. CCD technology has since been adapted to various types of imaging devices, including video and surveillance cameras. This technology has completely revolutionized astronomical image detection, both in the optical realm and for spectral analysis. CCDs provide numerous advantages over traditional photography and even other types of electronic cameras. CCDs are not only more efficient photon collectors, but also respond in linear fashion to the number and intensity of photons

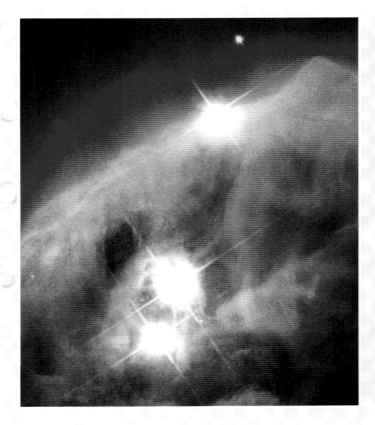

● Here are two Hubble Space Telescope images of a small portion of the Cone Nebula. The visible light image at left was taken with the ACS at 350 and 650 nm. The infrared image on the right was taken with the NICMOS camera at 2.2 and 3.5 μm. The 50 arcsec field shown represents about 0.6 light-years at the distance of the nebula, 2,500 light-years from us. Image resolution is approximately 0.1 arcsec, a mere 10 light-hours!

striking them. They have very low background noise and a broad dynamic range, thereby detecting both very bright and faint signals at the same time. CCDs are simple to operate but perform best when cooled to low temperatures around 100 K (–173°C).

During the 1970s, CCD chips typically contained only a few hundred pixels, but improvements over the years have increased that to over 10 million pixels today. This was achieved by linking several chips into larger arrays,

to the point where they have reached the size of the old photographic plates. CCD performance characteristics have also improved over the years, particularly with respect to dark current and internal noise, with markedly increased efficiency through image intensifiers. Improvements like these had immediate application in astronomy, particularly with respect to detection of very faint objects. Additional advances also made it possible to use CCD technology for observations at different wavelengths. For example, thinner

Major features and characteristics of charge-coupled devices (CCDs)

Dark current: the rate of generation of electrons by the sensor in the absence of light. It is due to thermal vibrations randomly exciting electrons.

Dynamic range: the ratio of the saturation level to the dark current. The wider this range, the better the detector is able to produce photometric measurements for both very faint and very bright objects in the same image.

Linearity: the ability of the detector to produce a signal in direct proportion to the number of photons it receives.

Quantum efficiency: the percentage of incident photons that generate photoelectrons. This is the ratio between the number of photons detected and the number of photons entering the detector.

Readout noise: random variations in electrical charge. These parasitic fluctuations add arithmetically to the output signal of the detector.

Saturation: a region above the linear range of the detector, in which the signal output is constant, irrespective of the input.

Spectral range: the range of wavelengths to which the detector is sensitive.

Spectral response: the sensitivity of the detector to different wavelengths of light.

Time constant: the time required by the output of a circuit to reach a percentage (63 percent) of peak value following a change in input.

● The dark nebulosity Barnard 33, popularly known as the "Horsehead Nebula," is shown above. This is actually a cloud of interstellar dust located 1,400 light-years from us in the constellation of Orion. This spectacular image was taken with the 8.2 m Kuyen telescope and its combined spectrograph-imager FORS 2. The image is a 17-minute composite exposure at wavelengths of 429, 554 and 655 nm.

chips and back-illumination improved performance at short wavelengths, and changes in semiconductor composition lessened impurities in chips and improved the detection of infrared radiation.

INFRARED DETECTORS

The application of infrared CCD detectors is a relatively recent development. A number of attempts were made during the 1970s to design infrared CCDs along the same lines as those used for visible light, but with a different type of semiconductor. Various types of materials were tried as a substitute for silicon without success, until so-called "hybrid technologies" were developed. This was achieved by combining photo-sensitive materials sensitive to the desired wavelengths with regular, silicon CCD chips. Such hybrid receptors have been perfected in recent years to the point where their performance is almost on par with visible light CCDs. Infrared detectors with a million pixels have been developed, almost as large as those for optical wavelengths. They span the range of 1 to 25 μm, beyond which other types of detectors (specifically bolometers) must be used.

DETECTORS OF THE FUTURE

The principal improvement currently sought for CCDs is the reduction of dark current and other electronic noise to as close to zero as possible in order to attain greater photon counting efficiency. Efforts are also underway both in Europe and the United States to develop entirely new types of detectors. In the future, superconducting materials will likely be used in low-noise detectors covering the X-ray to infrared range, and it is expected that there will be improved performance capabilities in spectroscopy. ■

■ *The Dunn Solar Telescope (DST) ranks among the most impressive astronomical instruments ever built. The Sun's rays enter the DST atop a 42 m high tower. The entire telescope measures nearly 110 m from top to bottom, with the principal components located under the ground.*

The Sacramento Peak Observatory

Imagine a theater presentation of lights and shadows, where silhouettes of pine trees appear to guard a huge tower rising from the Sacramento forest like a totem pole. Up here, the familiar constellations bear names given to them long ago by Native Americans. At night the Big Bear and its smaller companion circle the Pole Star amid a myriad of stars in the crystal-clear mountain air. Just a few steps from the tower we can see two hermetically sealed domes and, beyond that, a quiet building where astronomers are now fast asleep. The glow of lights from El Paso near the Mexican border lies to the south, and to the west we see an occasional truck along Highway 70, which borders White Sands National Monument. And then, just before dawn breaks, brilliant Venus appears above the tree line.

Perched at an elevation of 2,800 m, astronomers at Sacramento Peak Observatory actually observe only one star, the one closest to us, the Sun. The three domes of "Sac Peak," as it is known, only open after the Sun has risen high enough above the horizon to be clear of the densest parts of the atmosphere. The smallest of the three domes, nicknamed Hilltop, constantly monitors solar activity. The second houses an equatorially mounted spectroheliograph and a coronagraph with a .4 m objective lens. The last, an impressive 42 m high conical tower dominating the countryside, is one of the most powerful solar telescopes in the world. Much like an iceberg, the Dunn Solar Telescope (DST) conceals much of its bulk, as the base of its drive (which is 100 m high and weighs more than 250 metric tons) is located 65 m underground! This strange vertical monument was designed by a group of scientists, fully aware of the effects of the Sun's heat on telescopes and the need to mitigate this as much as possible, particularly in the summer. They realized that the air currents affecting optical performance

■ *The largest coronagraph in the world is located at an elevation of 2,800 m at Sacramento Peak Observatory. With its .4 m aperture objective lens, this instrument is dedicated to monitoring the Sun's chromosphere.*

would be most stable near the top of the DST, and that excessive heating along the telescope's optical path could be avoided by keeping the entire instrument sealed and under vacuum.

The rays of the Sun enter at the top of the tower through a movable heliostat containing a .76 m quartz window and two 1.1 m plane mirrors. The latter reflect the light down toward the telescope's 1.6 m parabolic primary mirror, located deep underground. This mirror focuses the light back to ground level where a .51 m diameter image of the Sun is formed. The instrument's focal length is 58 m. Since the resolving power of a telescope is a function of its aperture, the effective aperture of the DST is .76 m, or that of its quartz window.

Sac Peak astronomers are analyzing the Sun's image both photographically and spectroscopically. As with solar observatories worldwide, Sac Peak astronomers have tried for decades to obtain really high-resolution images of the Sun in order to investigate rapid changes in its atmospheric layers. Unfortunately, with the exception of a few remarkably calm periods, seeing conditions at the site average only 1 arc second.

To overcome this perennial problem, astronomers took advantage of recent developments in electronics and optics and equipped the DST with a system of adaptive optics that effectively eliminates the impact of atmospheric turbulence. Although this has narrowed their usable field of view to about 20 arcsec, resolution is now an impressive 0.15 arcsec, enough to reveal solar detail only a few hundred kilometers across. However, even higher resolving power is needed for more detailed studies of solar atmospheric dynamics and interactions with our star's powerful magnetic field. This has prompted plans for a huge 4 m telescope at Sac Peak to be known as the Advanced Technology Solar Telescope. ■

The Apache Point Observatory

The domes of Apache Point Observatory sit near the edge of a 2,800 m high crest in the Sacramento Mountains. Far to the west and much lower down, the dunes of White Sands glitter in the sunlight. The winter sports town of Cloudcroft is about 30 km to the north, and Sacramento Peak Observatory lies just a few kilometers away in the forest. Apache Point was only recently established as an astronomical site, as the decision to install instrumentation there was only made in 1984. The observatory was built and is managed by ARC, the Astrophysical Research Consortium, a group of American universities that now share time on the three telescopes at the site. The smallest instrument is only 1 m in aperture and is used primarily by astronomy students from the University of New Mexico.

When the observatory's largest instrument, a 3.5 m telescope, was placed into service in 1994, it was widely seen as the precursor to future generations of similar types of instruments. Prior to this it had not seemed feasible that a 3.5 m telescope could be so small, light and compact. With a focal length of only 6.1 m, its mirror had an f-ratio of only 1.75, a record at the time. In addition, with its open mirror configuration, the entire telescope is less than 8 m long and weighs less than 40 metric tons. In contrast, similar-size telescopes built between 1970 and 1980 weighed from 200 to 250 metric tons. The 3.5 m ARC telescope is equipped with CCD cameras and spectrographs intended for a wide range of studies, including solar-system asteroids and distant supernovae. However, what made this telescope so unusual at the time was its mode of operation: namely, remote control by any one of the six member universities of the ARC. In short, astronomers no longer had to travel to Apache Point to observe,

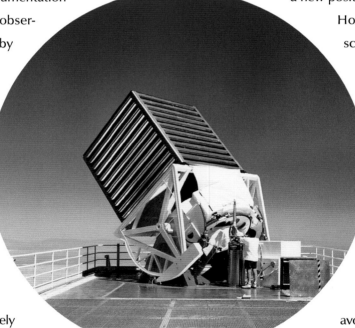

■ *The Sloan Digital Sky Survey's 2.5 m telescope is shown here. This instrument will survey an area of more than 10,000 square degrees between now and 2006, fully one-quarter of the celestial sphere. It is estimated that some hundred million stars and galaxies will be surveyed at Apache Peak in this manner.*

but could now access the telescope via the Internet. It is now used in this manner almost 75 percent of the time. Since it is so fast and easy to use, the 3.5 m telescope is also unusually flexible and can respond quickly when required for special deployment. For example, with the sudden appearance of a nova or supernova, whose early stages are often the most crucial, this telescope can readily interrupt its ongoing program and quickly slew to a new position in the sky.

However, a far less powerful telescope at Apache Point Observatory, a 2.5 m instrument, has drawn the most attention. This telescope is dedicated to one of the most ambitious sky mapping efforts ever attempted, the Sloan Digital Sky Survey, the modern equivalent of the famed photographic Sky Survey undertaken with the Schmidt telescope at Mount Palomar 50 years ago. This time, however, the survey will initially avoid regions of sky in and around the Milky Way, where star density is too high. The Sloan project, which began in 2000 and is slated for completion in 2005 or 2006, will focus on the galaxy and quasar-rich regions of the sky around the North galactic pole. Upon completion, the survey will have covered an area of 10,000 square degrees: one-quarter of the celestial sphere. Never before has a project like this involved such powerful image-detection technology. The 2.5 m telescope is equipped with a mosaic of 30 CCDs with 2,048 x 2,048 pixels each, covering 0.4 arcsec per pixel and a total field of 4 square degrees. The sky will be surveyed through five narrowband pass filters, each covering from 50 to 150 nm, centered at 354, 477, 623, 762 and 913 nm respectively. This spans the electromagnetic spectrum from the near ultraviolet, through the visible range to near-infrared wavelengths.

■ *The dim object indicated above (arrow), is a quasar discovered with the 2.5 m telescope at Apache Point Observatory. Subsequent analysis by the Keck I telescope in Hawaii revealed a red shift of z = 5.8, making it the most distant object observed to date, at a distance of more than 14 billion light-years. The Sloan survey should reveal about one million quasars.*

It is expected that an average of 20,000 objects per square degree will be recorded, down to 21–22 magnitudes. In total, the Sloan Digital Sky Survey is expected to map more than 100 million stars, several hundred million galaxies and about one million quasars. The final database will contain, both literally and figuratively, an astronomically large amount of information, more than 10 terabytes in total, all stored on magnetic tape.

A MILLION QUASARS

In addition to its purely photometric work, the Sloan survey will also provide spectra of many of the objects observed, using 640 optical fibers to record individual spectra with a resolving power of 2,000. One million galaxies and 100,000 quasars will be analyzed in this way. In order to identify the comparatively rare quasars (estimated around 50 per square degree) from thousands of other objects in the field, each survey image will be computer-processed to select point sources that are highly redshifted. This task will be made easier because all objects in the survey are automatically subjected to photometric analysis. Most quasars are readily identified because the ultraviolet portions of their spectra are shifted into the infrared, a direct reflection of the expansion of the Universe and their enormous distances from us. As the redshifts in the spectra of galaxies and quasars are a measure of distance, with each 0.2 displacement corresponding to about 2 billion light-years, the Sloan survey will soon provide us with a three-dimensional map of the Universe. As of 2002, the survey had already identified more than 50,000 quasars. With redshift in excess of 5, most of these objects are clearly at enormous distances in space/time, and of great interest to cosmologists. Several large telescopes, including the 3.5 m ARC, the 9.2 m at McDonald Observatory and the Keck telescopes, have all taken high-resolution spectra of the most distant quasars. This led to discovery by Keck I of the most distant quasar yet. With a redshift of z = 5.8, or roughly 90 percent back to the beginning of the Universe, this object appears to us as it did 14 billion years ago, or about 1 billion years after the big bang. When completed, the Sloan Digital Sky Survey will help define the space/time distribution of quasars and follow their evolution at various points in the Universe. It will also finally allow us to determine when the first galaxies were born. ■

The Mount Hopkins Observatory

UNITED
STATES

We are on a sharply rising mountain not far from the Mexican border between Tucson and Nogales. At 2,600 m above sea level, Mount Hopkins' summit lies well above the dusty Sonoran Desert and enjoys clear, steady skies. As they climb the tortuous road to the summit for the first time, it's difficult for visitors to imagine that a huge telescope is located up there. Nevertheless, the 6.5 m MMT crowns this precipitous mountain on an improbably small platform at its very top. The MMT is one of the strangest adventures in modern astronomy. In the 1970s astronomers had reached a major technological impasse. The biggest telescopes at the time weighed between 300 and 400 metric tons and were supported on costly equatorial mounts. Their large 3.6 to 5 m diameter mirrors were not only hard to equilibrate thermally, but tended to warp under their own weights. Consequently, for both financial and technical reasons it seemed unlikely that much larger telescopes would ever be built. While the Russians took their chances with the 6 m Zelentchouk telescope, in the United States, engineers at the University of Arizona and the Smithsonian Institute experimented with a variety of alternative telescope designs. This resulted in the development of the Multi Mirror Telescope (MMT) in 1980. This unusual and very futuristic instrument was not housed in a traditional, solid dome, but in a much simpler, movable building. Also, unlike most telescopes up to then, it was installed on a lighter, more compact altazimuth mounting. Its most innovative features, however, were its optics. These consisted of an array of six 1.8 m mirrors with an effective aperture equivalent to a single 4.5 m mirror. These features and the fact that this compact instrument weighed only 125 metric tons made the MMT a test case for technologies that have since become the norm. However, since it proved difficult to keep the

■ At an elevation of 2,600 m, Mount Hopkins Observatory is equipped with some of the most advanced instruments used today: the famed 6.5 m MMT, the 1.3 m project 2MASS telescope and the IOTA infrared interferometer.

optics in optimal alignment, the MMT was used primarily at infrared wavelengths and for spectroscopic work requiring high light-gathering power. Unfortunately, since technologies like adaptive optics and optical interferometry had not yet been developed, the telescopes effective 6.5 m aperture could not be used to full advantage for high-resolution work.

In the late 1980s, thanks to rapid advances in optical technologies, researchers at the University of Arizona decided to replace the six individual mirrors of the MMT with a single large one. This was accomplished quite quickly; a scant 15 years after it was first installed, the MMT was effectively gutted, its mount strengthened, and a new 6.5 m installed. The mirror fit the same dimension as the original instrument, but at f/1.25 it is a much faster optical system. The new instrument was able to utilize the same building and part of the old mounting, keeping the cost of the renovation at only about US $24 million. It even kept the same acronym, MMT, only now it stands for Monolithic Mirror Telescope.

The new telescope entered service in 2000 and offers astronomers an assortment of focal ratios: f/5.4, 9 and 15, corresponding to focal lengths of 35, 58 and 98 m respectively. With its all-Cassegrain configuration, the MMT supports several types of auxiliary instruments, including Hectospec, a spectrograph with 300 optical fibers that can be positioned anywhere in the telescope's field of view in just a few minutes. Other instrumentation includes Mirac 3, a 128 x 128 pixel infrared camera cooled to 5 K (−268°C) by liquid helium, and a widefield camera called MegaCam. The latter contains an array of 36 CCD chips, each with 4,608 x 2,048 pixels, for a total of 340 million pixels. Installed at the f/5.4 focus in 2004, this camera covers a field of view 24 x 24 arcmin, the largest to date for a telescope of this size. This combination will be used for deep surveys of the Universe down to magnitudes 28 and 29.

Although the actual summit of Mount Hopkins can only accommodate one major telescope, many of the mountain's 2,300 to 2,500 m high shoulders are dotted with observatory domes. These include a 1.2 m and a 1.5 m telescope, and a 10 m gamma-ray telescope, located just a few hundred meters from the MMT.

The final facility on the mountain, in operation since 1997, houses a 1.3 m telescope especially built for the Two Micron All Sky Survey (2MASS). This ambitious project is dedicated to an all-sky infrared survey and involves facilities at Mount Hopkins in the northern hemisphere and at Cerro Tololo Observatory in the southern hemisphere. Telescopes at both locations are equipped with 256 x 256 pixel array cameras that can record images at three different wavelengths: 1.25, 1.65 and 2.16 μm. In 2001 this equipment attained a record high-resolution of 2 arcmin in the infrared.

PROJECT IOTA

Mount Hopkins is also home to one of the first successful optical interferometers. Installed in 1988, the Infrared Optical Telescope Array (IOTA) was initially planned as an array of two .45 m Cassegrain telescopes on a mobile base about 15 m across. The consortium of five American universities involved wanted it to serve both as an astronomical instrument and a testing ground for new interferometric technologies. The first successful interference patterns were obtained in 1993 and the first scientifically useful results in 1995. IOTA was

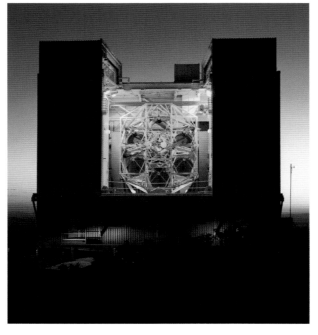

■ *When first installed in 1980, the MMT was a truly innovative telescope design, with its segmented primary optics, its altazimuth mount and its open dome. At that time, the MMT main mirror consisted of an array of six 1.8 m mirrors.*

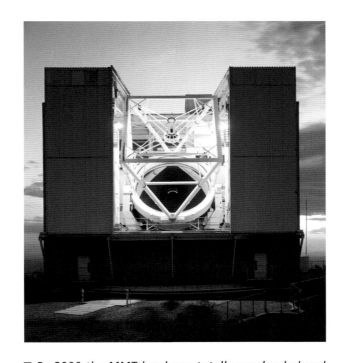

■ *By 2000 the MMT has been totally overhauled and equipped with a single mirror 6.5 m in diameter. Beginning in 2004, a new 340 million pixel CCD camera known as MegaCam will provide 24 x 24 arcmin field of view coverage, a record for such a large telescope.*

able to measure the diameters of the two red giants Mira Ceti and R Leonis with great precision, as well as those of several Cepheid variables. For example, the apparent diameter of the star RS Cnc was determined as 0.0156 arcsec, to an accuracy of about 0.001 arcsec, and another giant star RX Boo as 0.019 arcsec. Observing at both optical and infrared wavelengths, IOTA is equipped with several advanced technologies, including optical fibers and a "tip-tilt" mirror system that helps mitigate the effects of atmospheric turbulence. Though relatively simple, this system is very effective and very much in demand. Researchers from many other institutions have used it, including NASA Ames Research Center, Smithsonian Astrophysical Observatory, Harvard University and the University of Massachusetts. European collaborators include Paris and Grenoble observatories, the Astrophysical Institute of the Canaries and even the European Southern Observatory.

In 2002, a third .45 m telescope was installed at the site to form an L-shaped, mobile baseline with two 15 x 38 m branches. The maximum resolution attainable with this array now exceeds 0.01 arcsec at a wavelength of 165 μm, or ten times better than the Hubble Space Telescope. In practical terms that means that IOTA should separate the components of a binary star system 200 light-years from us, even if they were separated by only 100 million km. Such outstanding performance comes at a price, however. Since the total light-gathering power of this synthetic telescope is rather modest, it can only be used for stars bright enough to be seen with the naked eye. ■

> Widefield imaging

New developments in widefield imaging have changed traditional astronomical observing methods. Thanks to enormous advances in electronics and image processing techniques, this area of astronomy has broadened in scope in ways not even contemplated just a few years ago.

THE IMPORTANCE OF WIDEFIELD IMAGING

Just a few years ago a typical CCD camera on a 4 m telescope could cover a field just a few arc minutes wide at best, which was often too small to fully cover many star clusters and nearby galaxies. Consequently, several overlapping exposures and extended telescope time was needed for this. A narrow field of view is also not well suited for survey work in such areas as dark matter distribution, extrasolar planets, deep-sky probes, and small bodies in our Solar System. Worse still, these limitations made it almost impossible to conduct widefield mapping studies of faint stars and galaxies. Although extensive sky surveys had been carried out photographically for several decades, these were limited in several respects. Photographic surveys were only done at visible wavelengths, recorded only brighter objects, and were limited in angular and photometric resolution.

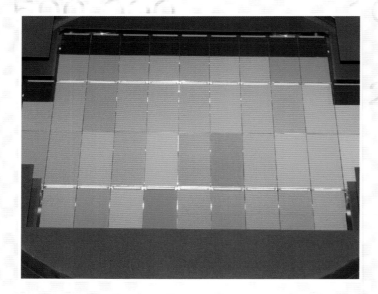

● Modern widefield cameras consist of arrays of several electronic receivers. The largest such camera currently in use, MegaCam, forms part of MegaPrime, and consists of 40 CCDs and a total of 360 million pixels.

The challenge was how to combine the widefield capabilities of photographic plates and the advantages of CCD technologies. Typically, plates measured tens of centimeters in length and width, ten times larger than any CCD chip at that time. Only two solutions seemed obvious: construct ever-larger CCD chips, or combine several smaller ones into a larger array.

Just as the size of telescopes has increased over the years, CCD cameras today contain hundreds more pixels than their "ancestral" versions in the late 1970s, and this number is likely to increase significantly in the future. Likewise the number of pixels in infrared detectors has increased to the point of matching that of CCDs, although that is still not enough.

DETECTOR ARRAYS

Research on dark matter in the 1990s prompted the development of the first CCD arrays in France with EROS and in the United States with MACHOS. Thanks to great advances in digital imaging, electronics and information processing, it became possible to combine several CCD chips into single arrays for widefield astronomical work. In 1998 the 2.2 m telescope at La Silla Observatory was equipped with a Wide Field Imager consisting of eight CCD chips covering a field 0.5 square degrees, and in 1999 the CFH12K camera was installed on the Canada–France–Hawaii Telescope (CFHT). This camera, an array of 12 CCDs of 2,024 x 4,048 pixels (or simply 2K x 4K) each and a 0.3 square degrees field, led directly to construction of MegaPrime, the largest imager ever built. Composed of 40 CCDs of 2K x 4.5K pixels each, it covers a field of 1.4 square degrees, several hundred times more than a single chip.

WIDEFIELD RESEARCH PROJECTS

Thanks to the foregoing developments, several widefield imaging projects are currently underway involving the MMT

● This is a picture of the galaxy NGC 4945 in the constellation of Centaurus, taken by the 2.2 m telescope at La Silla Observatory with its Wide Field Imager (WFI). This camera, with nearly 68 million pixels, covers a 32 x 32 arcmin field of view (larger than the angular dimensions of the full Moon). More than 70 billion photons passing across three different filters were collected during the three-hour exposure needed for this image.

at Mount Hopkins Observatory, the Gemini and Subaru telescopes on Mauna Kea, Mount Wilson Observatory and several other dedicated instruments. In the northern hemisphere, the five-year Sloan Digital Sky Survey uses a 2.5 m telescope with a thirty 2K x 4K CCD mosaic to map the position and magnitude of several hundred million objects. A similar survey is planned for the southern hemisphere, using the VLT Survey Telescope equipped with a thirty-two 2K x 4K CCD detector for a 1-square-degree field. Even more ambitious, project VISTA in 2004 will use the 4 m telescope at Cerro Paranal and a 2.2-square-degree field.

Much progress has also been made in the development of infrared imagers, notably the WIRCAM camera for the CFHT, covering a 0.16-square-degree field, and VISTA, with a 1-square-degree field and nine 2K x 2K infrared detectors. Only 20 years ago all infrared-sensitive sensors were effectively "monopixel" devices, which illustrates just how much progress has been made in this field.

This drive toward wider field imaging has not merely been a race to build bigger instruments. Rather, it has been a concerted 10-year effort to explore new possibilities by astronomers with increasingly complex needs and by electronic experts seeking improved CCDs. The results have opened new horizons in all areas of astronomical research. These range from the detection of small objects in the Solar System to surveys of highly redshifted objects in the Universe, including galaxies, clusters

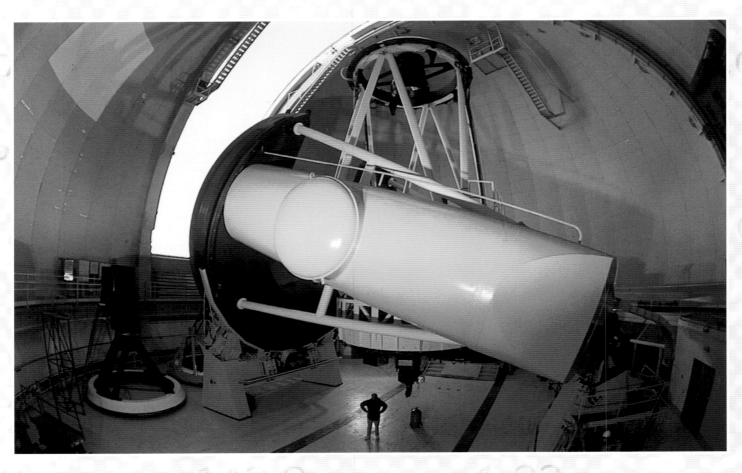

● In the coming years the 3.6 m Canada–France–Hawaii Telescope will specialize in widefield work. Its MegaCam camera will cover a field 1°10' x 1°10' at the prime focus, seen here at the top of the telescope's tube. This camera should reach 26th magnitude during a 30-minute exposure. The CFHT will explore several hundred square degrees of sky in search of asteroids, brown dwarfs and supernovae in distant galaxies.

of galaxies and associated gravitational lensing effects. New technologies have also led to advances in stellar physics and the search for extrasolar planets.

● Weighing some 300 kg, MegaCam's dimensions are 1.7 x 1.2 m. It contains an array of CCD chips measuring 31 x .26 m, which will generate 770 megabytes of data per exposure.

A TRIPLE CHALLENGE

The successful development of widefield electronic cameras faced several major challenges. Assembling large arrays of detectors was not a simple process and required a number of technical innovations. It was important to avoid cross-reaction among the many chips in these mosaics and to keep signal downloading as short as possible to minimize noise. To do this, several CCDs must be read simultaneously. The 40 chips in the CFHT MegaCam camera are read in about 20 seconds, which is only four times longer than required for a single chip, during which less than 10 electrons of internal noise are generated.

Another challenge with any telescope is that image quality and signal strength fall off sharply the farther off axis you go. This is due to coma and other optical distortions. To avoid this, field flatteners must be placed in front of cameras, particularly for widefield applications, to ensure uniform

● M33 is a spiral galaxy located some three million light-years from us in the constellation of Triangulum. Individual stars in the arms of the galaxy are fully resolved in this CFHT image, taken with the CFH12K camera. Most of the stars are blue supergiants: hundreds and thousands of times brighter than the Sun. Numerous nebulae are also visible across the entire galaxy, ionized by radiation clusters of newborn stars.

image quality across the full optical plane. Image stabilizers are also used to help keep images as sharp as possible. Lastly, it is important that star images are registered as sharp points across the entire field. The MegaPrime camera excels in this regard. Fully 80 percent of the energy in stellar diffraction patterns falls within a 0.3 arcsec diameter disk, and this across the entire 1.4-square-degree field of view!

The final challenge concerns data storage and handling. For example, at the CFHT, MegaPrime with its 350 million pixels records 700 megabytes per exposure, and generates several dozen gigabytes of data during a typical observing run. Likewise, the Sloan Digital Sky Survey generates more than 40 terabytes of data per night! Data collection, handling and storage of such magnitude are a relatively new development in astronomy, and new techniques had to be developed to deal with it. Consequently, supercomputers and data storage facilities have been established at several institutions, similar to those already used by space-based instruments. ∎

The Starfire Optical Range

Secluded in the high desert of New Mexico and girdled by Kirkland Air Force Base, the Starfire Optical Range (SOR) sits atop a small hill covered with domes, radar antenna and a ring of electrical wiring. Unknown to the larger astronomical community prior to 1991, this mysterious observatory is still engaged in military surveillance work using cutting-edge technologies.

UNITED STATES

The cold night has descended and the coyotes and prairie dogs are below ground. One after another, a series of aluminum domes opens up, covered only by reflected starlight. An unusual spectacle now takes place on this remote, 2,000 m high desert plateau. An intense column of green light suddenly appears in the sky, seemingly pointed at one or other of the domes. Since nothing was visible just a moment ago, it's hard to tell whether the laser light came from the sky or the ground. Yet now, like a scene out of *Star Wars*, an immense sword of light points rigidly skyward. More surprising still, the beam does not point at any stars, but seems to end quite abruptly high in the sky.

Experimental work with lasers has been going on under New Mexico skies for at least 20 years, yet almost no one ever gets to see this. The Starfire Optical Range is well hidden amid the high mountains, but close to other secret government labs not far from Kirkland Air Force base, near Albuquerque.

This facility has probably been around since 1978. In 1980 it became part of the Strategic Defense Initiative (SDI) or "Star Wars" under President Ronald Reagan, who sought to develop a variety of powerful laser weapons capable of destroying incoming Soviet nuclear missiles. However, like astronomers before them, Pentagon researchers soon came up against a perennial problem: atmospheric turbulence. Almost limitless funds were directed at

■ *Equipped with powerful lasers and adaptive optics, telescopes at the Starfire Optical Range are well hidden at Kirkland Air Force Base. The existence of these facilities was not acknowledged prior to 1991.*

this predicament, and the ultimate result was the development of adaptive optics.

The U.S. government has experimented with lasers for a variety of military purposes. Laser-equipped telescopes were used for many years to target spy satellites and incoming missiles with highly coherent beams of light. These telescopes also send sharply focused laser light high into the atmosphere, where it interacts with air molecules at altitudes of 10 to 40 km, and causes ionization in the sodium layer around 90 km. Seen from Earth, the resultant points of light resemble artificial stars following any target the telescope is directed at. These images are used to monitor atmospheric turbulence, which can then be corrected with adaptive optics.

In 1983 when the first tests were being conducted at the highly classified Starfire Optical Range, two French astronomers, Renaud Foy and Antoine Lebeyrie, published a short article in the European journal *Astronomy & Astrophysics*, outlining the theoretical basis for an adaptive optics system using a laser-projected artificial star as a reference point. Most astronomers considered this highly unusual idea unrealistic at the time, and, since no one had the means to test it, it was not followed up.

The U.S. military's dream of creating a total protective shell over the entire country was never realized. SDI was eventually abandoned and much of the research was declassified. However, the billions of dollars invested in the technology had peaceful applications in observational astronomy. During the 1990s, unlike other classified facilities, this observatory welcomed a handful of scientists from all over the world interested in high-resolution work, and provided access to its smaller 1.5 m telescope while construction of a civilian 3.5 m telescope with adaptive optics was underway.

The 1.5 m telescope is equipped with a laser emitting at a

wavelength of 570 nm and a pulse rate of 5,000 times per second. Although each pulse lasts only 50 nanoseconds, it emits almost 1 megawatt of energy and projects a brilliant artificial star high in the atmosphere. The image is constantly monitored and corrected by the telescope's adaptive optics, which consist of 208 microlens filters and a flexible mirror that can be deformed 130 times per second through 241 actuators. A team led by Robert Fulgate used this facility primarily to investigate bright, point-source objects, such as double stars and asteroids.

AN ULTRAFAST TELESCOPE

Beta Delphini is a beautiful, very close and exceptionally rapid double star system. Thanks to adaptive optics, the 1.5 m telescope was able to resolve it, even when the two components were only 0.20 arscec apart. In similar fashion, the shapes and exact dimension of three asteroids, Vesta, Ceres and Pallas, were accurately determined for the first time. Unlike Vesta and Ceres (the two largest asteroids in the Solar System), Pallas appeared slightly elliptical in shape rather than perfectly spherical.

These days, astronomers are working with a new 3.5 m telescope at this facility. Set on a mount that can slew at 12° per second, this large instrument can sweep the sky from horizon to horizon in only 15 seconds! In short, it can follow any rapidly moving spacecraft, even one in low orbit. Since such rapid motion is beyond the capabilities of traditional observatory domes, the Air Force has housed this telescope in a very unusual fashion: in a series of interlocking cylinders that are automatically lowered at twilight, leaving the telescope in open air. This greatly lowers any air turbulence in and around the telescope and takes full advantage of its still-classified system of adaptive optics. Among many other non-military applications, this powerful telescope will provide NASA with new laser-based technologies to communicate with the next generation of space probes.

With adaptive optics, the performance of the 3.5 m SOR telescope is comparable to that of existing civilian instruments similarly equipped, including the CFHT in Hawaii and the 3.6 m European telescope at La Silla. However, with the advent of new giant telescopes like the VLT, Keck II, Gemini and Subaru, which are also equipped with adaptive optics, the Starfire Optical Range instruments will soon become obsolete. Of course, it's highly unlikely that the U.S. Air Force will let that happen. Is it possible that an ultrasecret, 8, 10 or 12 m telescope is already hidden somewhere in the New Mexico desert? ■

The Mount Graham International Observatory

UNITED STATES

Amid century-old spruce trees covered with countless beautiful red cones, with a crystalline blue sky above, the summit of Mount Graham is one of the last pristine astronomical sites in the continental United States. This imposing mountain in the middle of the Arizona desert is located over 120 km north of Tucson and tops out near 3,200 m above sea level. The thickly wooded mountain was only recently recognized as a premier astronomical site. The Submillimeter Telescope Observatory (SMTO) is a joint German and American venture. The first telescope, placed there in 1994, was the 10 m submillimeter radio telescope. Observing from 0.3 to 1 mm, this technological marvel was designed to fill the gap between infrared and radio wavelengths. Its antenna, made of carbon fiber material, has a surface accuracy better than 15 μm. The submillimeter window, at the fringes of optical and radio technologies, was accessed only a few years ago and so remains largely unexplored. The SMTO lets astronomers explore the coldest regions of the cosmos, including interstellar clouds, dark nebulae and protoplanetary disks around young stars.

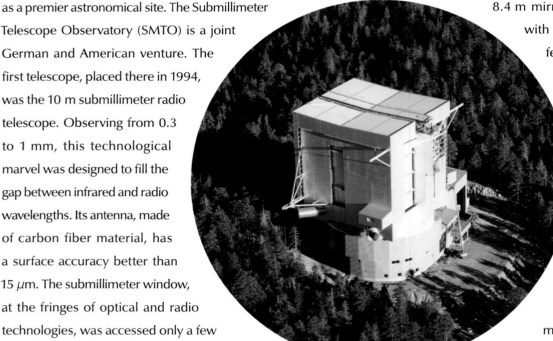

■ *At an elevation of 3,170 m, the summit of Mount Graham is one of the best astronomical sites in the continental United States. The 40 m high structure shown here houses the Large Binocular Telescope, which will be fully operational in 2005.*

In 1995 the SMTO was joined by another high-technology facility, the Vatican Advanced Technology Telescope, or VATT. It is managed by the Vatican Observatory at Castel Gandolfo, which was founded by the Jesuits in the 19th century under Pope Leo XIII. Vatican astronomers today have access to this compact, 1.8 m Mount Graham telescope, which has an extraordinarily fast f/1 focal ratio. This configuration lets VATT researchers capture images of nebulae and distant galaxies with exposures of just a few seconds or minutes.

The third major telescope on Mount Graham is currently in its testing stages and slated for full operation in 2005. The Large Binocular Telescope (LBT) is one of the most ambitious astronomical projects of the early 21st century, and on par with other large optical interferometer efforts like the Keck telescopes and the VLTI. The design characteristics of this new type of instrument are simple yet quite spectacular: just as its name indicates, it will be a gigantic pair of binoculars! The LBT's huge altazimuth mount weighs around 380 metric tons and supports *two* 8.4 m mirrors separated by a 6 m space. As with regular telescopes, each mirror will feed a suite of auxiliary instruments, but the primary function of the LBT will be to observe with both mirrors concurrently, bringing their respective light paths to a common focus. In that configuration, the LBT will have the resolving power of a 22.8 m aperture telescope. In principle this level of resolution will only be achieved in the wider axis of separation between the two mirrors; the other axis will be limited to 8.4 m, or the diameter of each mirror. This can be readily compensated for, however, by observing the same object at the beginning, middle and end of the night. By taking advantage of the apparent rotation of celestial objects as they move across the sky, the LBT will face them from all angles, including at its widest effective aperture of 22.8 m. Subsequent image processing will take care of the rest.

To help finance a technological and scientific project of this size, an international consortium of fifteen universities and national research institutes are involved. The United States has provided 50 percent of the estimated $90 million needed for the LBT, with Italy and Germany contributing 25 percent each. The telescope's enormous mount was built in Italy and installed in the 40 m high dome in 2002. The two borosilicate glass mirrors were ground and polished at the University of Arizona's Mirror Lab in Tucson. They were

■ *This is one of the most ambitious and innovative astronomical instrumentation projects ever undertaken. Equipped with two 8.4 m diameter mirrors on a single mount, the Large Binocular Telescope movable mass exceeds 380 metric tons. The participants in this international effort expect the giant binoculars to attain image resolution near 0.01 arcsec.*

cast there as well, in 1997 and 2000, using an extremely innovative method developed by American astronomer and optical engineer, Roger Angel. The borosilicate glass was heated to 1,150°C and melted in a rotating oven spinning approximately seven times per minute. To keep the mirror as lightweight as possible, the glass was molten onto a base of 1,600 heat-resistant tiles arranged in a honeycomb pattern. The centrifugal force generated through spinning helped shape the surface of the molten glass into a natural parabola. Several weeks later, after the mirror had sufficiently cooled and solidified, its surface was polished and finished. The two 8.4 m mirrors are the largest and most lightweight ever produced (only 16 metric tons each), and with a focal length of only 9.5 m, they are the fastest optical systems yet devised, with focal ratios of f/1.14. Prior to undertaking this huge project, Roger Angel's team learned its trade on several other projects, including the open configuration 1.8 m VATT and the 3.5 m mirrors for the telescopes at Apache Point and WIYN observatories. They also made three 6.5 m mirrors, one for the Mount Hopkins MMT and two for the Magellan project telescopes.

If everything goes as expected, the LBT's optical performance should be exceptional. Its effective aperture falls squarely between that of conventional telescopes in the 8 to 10 m class and optical interferometers in the 100 to 200 m aperture range. Unlike the latter, however, including the Keck and VLTI interferometers (which will be limited to observing brighter, point-source objects), the LBT will also be able to image extended objects. With its effective aperture of 22.8 m (corresponding to a single 11.8 m mirror), the LBT functions more like a conventional telescope in that it can directly photograph the planets, stars, nebulae and galaxies. It will operate both in the visible and near-infrared ranges, at wavelengths spanning 400 nm to 30 μm. In the interferometer mode, it should produce images three times sharper than an 8 m instrument and ten times sharper than the Hubble Space Telescope. When observing at 1 μm, the LBT should attain resolution of 0.01 arcsec. This is sufficient to detect, for example, individual volcanoes on Jupiter's moon Io, or objects just a few light-hours across in the core of our galaxy, near its massive black hole. Lastly, Mount Graham astronomers hope to be first to directly image an exoplanet. Given its anticipated superiority in both resolution and light-gathering power, the Large Binocular Telescope is likely to realize this goal, unless its competitors at Keck and VLTI reach it sooner. ■

The Mount Locke Observatory

From our vantage point atop Mount Fowlkes, to the south the venerable domes of McDonald Observatory on nearby Mount Locke glitter in the bright sunlight. From this perch some 2,000 m above sea level, we see before us an important chapter in the annals of American astronomy. The world's second largest telescope was built here in the 1930s by the universities of Texas and Chicago, thanks in large measure to a $1 million gift from William Johnson McDonald, a banker from Paris (Texas). When the 2.2 m Otto Struve telescope was placed in service here in 1939, it was second only to the famed 2.5 m Hooker telescope at Mount Wilson. Thirty years later, in 1969, an even larger telescope was installed on Mount Locke: the 2.7 m Harlan Smith. This instrument too was among the largest in the world for several years, surpassed only by the Hale telescope at Palomar and the 3 m Shane reflector at Lick Observatory. Following the wishes of its Texan benefactor, who saw astronomy as serving both science and culture, McDonald Observatory has always welcomed visitors and amateur astronomers. Today, in addition to nightly observing at its visitor center, the observatory has frequent public observing nights with both the Otto Struve and Harlan Smith telescopes. This is a unique treat for all participants since the skies there are among the best in the United States.

Mount Locke is located at the southernmost extension of the Rocky Mountains in one of the least populated areas of the country. This beautiful high desert location, between Pecos and El Paso near the Mexican border, enjoys more than 250 clear nights a year and gets only 330 mm of rain. The observatory's principal instruments were among the world's largest during the 1970s, but today rank 37th and 52nd respectively. While these

■ *The Hobby-Eberly telescope's segmented primary mirror array measures 10 x 11 m. Each of its 91 hexagonal mirror segments is 1 m in diameter, and the full mosaic is slanted at an angle of 55°. The total array can only turn in a vertical plane.*

are still widely used by the faculty and students at the University of Texas, the most important facilities at this site are now on Mount Locke. The 9.2 m Hobby-Eberly telescope, one of the most unusual modern astronomical instruments, has been there since 1997. Sitting amid very beautiful desert vegetation, its huge 30 m dome dominates the top of the mountain.

The design of this giant telescope is highly innovative. Built by a consortium of German and American universities, including the universities of Texas, Pennsylvania and Stanford in the United States, and Gottingen and Munich in Germany, researchers wanted a very powerful yet economical instrument. As it turned out, the project cost less than $15 million, or about 15 percent that of an 8 to 10 m conventional telescope. Unfortunately, this also necessitated some big operational compromises; the telescope cannot image anything but is solely used for spectroscopy.

The design of this instrument was inspired by the Arecibo radio telescope. Its primary mirror is mounted in a fixed position, but inclined at an angle of 55° above the horizon. Weighing about 100 metric tons, the entire assembly can only turn in azimuth or along the vertical axis of an altazimuth mount. This clearly limits its mobility and makes it impossible to track celestial objects across the sky. Like Arecibo, however, this telescope is also equipped with a very sophisticated, six-degrees-of-freedom guidance assembly at the prime focus, which allows it to track objects for up to one hour. Even with this unique tracking mechanism, however, the Hobby-Eberly telescope only has access to about 70 percent of the sky above Mount Locke.

The telescope's primary mirror array is slightly elliptical in shape, measures 10 x 11 m, and is kept in constant alignment via 273

■ *The Hobby-Eberly telescope sits atop Mount Fowlkes at an elevation of 2,070 m. Behind it, McDonald Observatory is visible here on the summit of Mount Locke. The latter has served both amateur and professional astronomers since 1939.*

separate actuators. Each of the 91 individual 1 m mirror segments is 50 mm thick. Because of its unique configuration, the tracker sees only a portion of the mirror while observing, thereby limiting the effective aperture of the Hobby-Eberly telescope to 9.2 m.

A South African twin

Between 1999 and 2002, the Hobby-Eberly telescope was outfitted with a series of increasingly higher-resolution spectrometers. The low-resolution spectrometer (LRS), 600 to 3,000, is mounted directly on the tracker at the instrument's primary focus. The 3,500 to 21,000 medium-resolution spectrometer (MRS) is located lower down in the building, as is the highest-resolution spectrometer (HRS). The latter provides resolution levels of 30,000, 60,000 and 120,000, at wavelengths ranging from 420 to 1,100 nm.

■ *Commissioned in 1997, the Hobby-Eberly telescope is lightweight and simple in design. Like SALT, its twin in Sutherland, South Africa, this instrument is dedicated exclusively to spectroscopic work.*

Due to its great light-gathering capacity, this telescope provides researchers with spectra of even relatively faint objects in a reasonably short time. With exposures of only one to two hours, excellent spectra can be obtained of stars, galaxies and quasars between 20th and 23rd magnitude. The telescope has been particularly helpful obtaining spectra of distant supernovae and potential quasar candidates, identified from images taken with conventional telescopes. The unusual design and affordability of this facility prompted Polish and South African astronomers to build one just like it for the South African Astronomical Observatory (SAAO) in Sutherland. That instrument, to be known as the Southern African Large Telescope (SALT), is slated for completion in 2005. This type of optical design has also been considered for the next generation of large telescopes in the 30, 50 and 100 m classes. ■

> Optical interferometry

Astronomers have worked hard the past few decades at improving equipment and developing new techniques to observe the stars. In the future it is likely that both ground- and space-based interferometry will provide the highest levels of resolution. Below is a brief review of interferometry at visible and infrared wavelengths.

THE LIMITATIONS OF REGULAR TELESCOPES

The size of telescopes has increased steadily since the days of Galileo; today the largest exceed 10 m in aperture. Over the past 400 years the resolving power of telescopes has increased from a few arc seconds to a few hundreds of arc seconds for the largest modern instruments with adaptive optics. This is likely to improve even more in the coming decades when the current generation of telescopes is replaced by even larger instruments with segmented mirrors 30 to 100 m in diameter. It is unlikely, however, that telescope sizes will increase much beyond that, since both optical and mechanical limitations are likely to make that impractical. Extremely large optical surfaces are likely to sag under their own weight, and polishing

● The KTI interferometer in Hawaii will soon be fully operational. It consists of the two 10 m Keck telescopes separated by a baseline of 80 m. Astronomers hope to eventually link all Mauna Kea telescopes into a single interferometer called OHANA with an 800 m baseline.

mirrors with surfaces accurate to a 10th of a wavelength over a truly huge area may not be possible. Hence the maximum practical level of resolution of traditional telescope designs has probably been reached.

THE EARLY STAGES OF INTERFEROMETRY IN ASTRONOMY

Shortly after Thomas Young's pioneering experiments in the late 19th century, several physicists (including Hippolyte Fizeau in France and Albert Michelson in the United States) realized the potential applications of interferometry in astronomy, particularly with respect to resolving close binaries and measuring the diameters of nearby stars. Although these early efforts successfully determined the actual diameters of several stars and satellites, they were technically so difficult that astronomers abandoned optical interferometry in the 1920s in favor of larger single-aperture telescopes.

Radio astronomers started experimenting with interferometry again in the 1950s and quickly perfected it for use at radio wavelengths. Similar developments did not take place in optical astronomy until the 1970s, when Antoine Lebeyrie revived it in France. Most major observatories today are gearing for visible and infrared interferometry, and plans are underway for space-based interferometry in the coming decades.

THE PRINCIPLES OF TWO-ELEMENT INTERFEROMETRY

Separate rays of light from a single source can interfere with each other in either a constructive or a destructive fashion, depending how far their waves are out of phase. In astronomy this is done by combining light from one object simultaneously observed by two or more telescopes. This

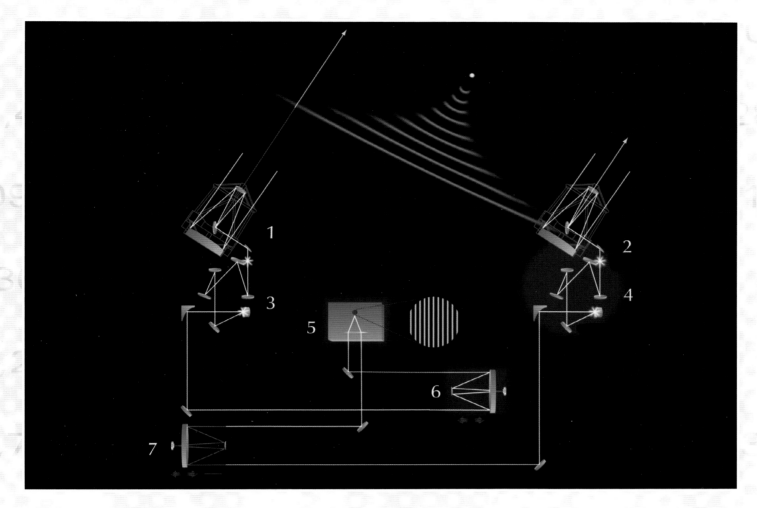

● In an optical interferometer light from the same object is collected simultaneously by two telescopes (1 and 2). The separate beams are then relayed by a series of mirrors (3 and 4) onto a single receptor (5), where they are combined into interference patterns. Since light reaching the two telescopes is slightly displaced due to the star's apparent motion in the sky, a series of movable mirrors called the "line of retardation" (6 and 7) bring the two beams back into register before they are recombined to form interference patterns.

produces a synthetic telescope whose angular resolution is equivalent to that of a single telescope with a diameter equal to the distance (or baseline) separating the individual components of the array.

TELESCOPE ARRAYS

Two telescopes working as an interferometer provide excellent angular resolution, but only in the direction separating the two instruments. This is adequate for point-source objects like stars, where unidirectional measurements are enough for diameter determinations. Measurements in more than one direction, however, are required for similar information on extended objects. This is accomplished by observing the same object several times in succession as its orientation changes relative to the telescopes' baseline. Known as "aperture synthesis," this technique can be used for both optical and radio interferometry. Though effective, this method needs a great deal of telescope time, sometimes several weeks or even months, before enough data is obtained to fully cover the target. Moreover, aperture synthesis is not effective for objects fluctuating in brightness if the sampling times needed are as long as the variation itself.

The rotation of the Earth can be utilized for interferometry based on directional differences. Known as "supersynthesis," this technique also requires extended telescope time and is not suitable for variable objects.

Multitelescope arrays go a long way toward overcoming this obstacle, since they can be used for concurrent target measurements from several directions at once. An array of three telescopes, for example, provides three separate baselines for concurrent measurements in three different directions; an array of six telescopes has 15 possible baselines, etc. As a general rule, an array of (n) telescopes provides n (n-1)/2 possible baselines per observing session. That is why multiple mirror arrays are highly desirable. For instance, the ESO's Very Large Telescope Interferometer (VLTI) will consist of four 8 m telescopes and several smaller, auxiliary instruments. Three of these have already been completed, with the rest likely to follow soon. The auxiliary telescopes are the most mobile of the array. With about 20 possible stations, they provide a large number of baselines and aperture synthesis in several directions. The Keck interferometer on Mauna Kea is designed along similar lines, with the two large 10 m telescopes and a battery of smaller, movable instruments called "outriggers."

● At present the Very Large Telescope Interferometer (VLTI) consists of two 8.2 m telescopes separated by a baseline of about 100 m. Once the four-telescope array plus the three smaller 1.8 m mobile instruments are ready, they will provide several baselines and form an interferometer about 200 m in diameter.

THE VARIOUS TYPES OF INTERFEROMETERS

It is not easy to generate interference patterns of astronomical objects, as this requires ultraprecise alignment of light beams from two or more telescopes. This can be done in two different ways: through Fizeau recombination or by the Michelson method. The first technique combines two or more beams directly to form an image whose quality and angular resolution will depend directly on the number of different baselines involved. If enough baselines are used then images equivalent to those generated by a telescope of comparable aperture is obtained. The Michelson method does not form actual images but generates interference patterns that are then analyzed mathematically to provide information on the size and dimensions of the object under study. With sufficient data it is also possible to reconstruct an image of the object.

● The main laboratory of the VLTI is located in the basement of the Cerro Paranal Observatory. The long tunnel constitutes the major baseline of the array. The VLTI will work at 0.6 μm in the visible range and at 20 μm in the infrared at three different foci: Amber, Midi and Prima. With its 8 m telescopes, the VLTI will reach 0.0036 or 0.00196 arcsec resolution at wavelengths of 2.2 μm and 1.2 μm respectively.

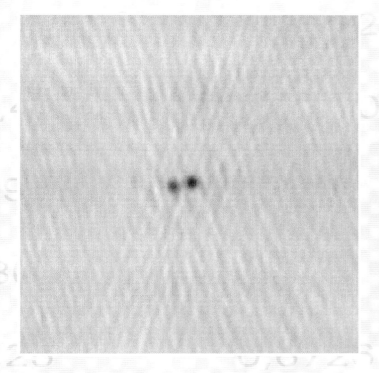

● This image of the double star Mizar was obtained with the optical interferometer NPOI at the Anderson Mesa observatory in Arizona. Six .5 m telescopes and a 60 m baseline were used to resolve this double star, separated by only 0.0065 arcsec. This degree of resolution is ten times better than can be achieved by the Hubble Space Telescope or the best existing adaptive optics.

● This illustration shows several dozen interferometric measurements of the star Achenar, taken with Antu and Melipal, two of the telescopes of the European VLTI array. These data set the apparent diameter of Achenar at 0.0019 arcsec, or an actual diameter of 13 million km for this star 145 light-years distant from us.

Since the Michelson-type interferometer is easier to operate with ground-based facilities, it was the method of choice for both the VLTI and the Keck interferometer.

THE KEY COMPONENTS OF INTERFEROMETERS

In order to generate useful interference patterns between two or more beams of light, they must be in perfect phase, which means their optical paths must be identical. However, atmospheric turbulence and slight differences in the angle of incident between receiving telescopes can result in phase shifts. This must be fully compensated for with residual errors not exceeding a fraction of the wavelength of the light in which the observations are made.

Adaptive optics is essential, especially with large aperture telescopes, to ensure that the wave fronts of all light beams are as uniform as possible. Differences in optical path lengths can be corrected directly: either by slight displacement of the telescopes themselves or, more simply, through various phase-displacement devices. Minor atmospheric instabilities or seeing effects can be corrected through fringe stabilizers, which even out interference pattern fluctuations for more uniform results. Such devices can be used directly on the object being observed or on a brighter reference star nearby. This is known as "twin field" interferometry.

If we had to use a single descriptive term to define interferometry it would be "precision." Precise control of optical path lengths, precise control of the distance between different elements in the interferometer, precision in actual measurements undertaken, and so forth. The margins for error with this technique are very small indeed: a fraction of the wavelengths of light studied, or a few dozen nanometers at most. It's not surprising in retrospect that Michelson's ideas took so long to be put into practice!

THE FUTURE OF INTERFEROMETRY

Ground-based optical interferometry has finally matured and thanks to this, we are likely to see numerous new discoveries in the coming decade. Space-based application is clearly the next phase in the development of this technology, which should provide excellent observational data, particularly at visible and infrared wavelengths. Exciting new projects in the planning stages include interferometric imaging of extrasolar planets and detection of surface detail not only on giant exoplanets but possibly on Earth-like planets as well. Obviously the building of such advanced interferometers will be gradual, and start with feasibility demonstrations of space-based instruments. Once there are some encouraging initial results, probably toward the end of the current decade, the real adventure will begin. ■

The Haleakala Observatory

The moon-like landscape is under perpetual sunlight and intense blue skies. Part of the Haleakala National Park, the Haleakala volcano provides a magnificent black and white panorama, with occasional ocher- and yellow-colored waves of solidified lava. Haleakala, "the house of the Sun" in Polynesian, is truly the domain of our own star. University of Hawaii astronomers did well to choose this site for a solar observatory, since they can observe from here more than 300 days a year.

The Hawaiian archipelago consists of eight principal islands formed roughly in sequence: Hawaii, Maui, Molokai, Lanai, Kahoolawe, Kauai, Niihau and finally Oahu, site of the state capital, Honolulu. The islands were formed one after the other, east to west, as the Pacific plate migrated over a "hot spot' in the Earth's crust. Most of the islands are huge, extinct volcanic domes, except for Hawaii itself, the largest, youngest and most easterly island of the archipelago, which is still volcanically active.

Lying about 100 km west of Hawaii, the island of Maui tops out at 3,050 m above sea level at the summit of Haleakala. The observatory is just a few meters below the summit, overlooking an immense caldera about 12 km long and nearly 1,000 m deep. The dome houses a powerful .25 m coronograph, a high-resolution spectrograph and a photopolarimeter, all clustered on the same equatorial mounting. At an elevation of more than 3,000 m, the site is well above the lower third of the Earth's atmosphere, where most water vapor and dust is concentrated. The mountain is typically ringed at 2,500 m by an inversion layer of ocean-driven clouds, with exceptionally pure skies at the summit: ideal conditions for

■ *Astronomers at Haleakala Observatory can monitor the Sun more than 300 days a year. At 3,050 m above sea level, the atmosphere atop this extinct volcano is exceptionally transparent and ideally suited for observations of the solar corona.*

solar observing. The observatory's primary mission is to monitor solar activity. In this regard, its location in the middle of the Pacific Ocean provides an ideal complement to similar activities at Kitt Peak Observatory in Arizona, Sacramento Peak in New Mexico, Pic du Midi in the Pyrenees, and La Palma and Pico de Teide in the Canary Islands. Distributed around the globe, these facilities provide round-the-clock coverage of solar activity.

SATELLITE TRACKING

Just a few steps from the solar observatory we find the Maui Space Surveillance Site (MSSS), originally a supersecret U.S. defense department facility, but now more and more accessible to civilian scientists. Just like the renowned Starfire Optical Range in New Mexico, scientists and engineers at MSSS have developed optical and electronic technologies to identify and monitor all Earth-orbiting satellites. Before astronomers gained access to and installed similar technologies on virtually all modern large telescopes, the U.S. military developed adaptive optics to minimize atmospheric interference during high-resolution monitoring of space vehicles and debris over the Pacific Ocean. Initially this involved several "small" telescopes at MSSS, ranging in aperture from .9 m to 1.2 and 1.6 m, but ultimately it led to development of a facility of great interest to the military and astronomers alike. The Advanced Electro-Optical System (AEOS) telescope is officially the most powerful optical instrument in the Department of Defense the arsenal. Placed into operation in 2000, this instrument has occasionally been made available to astronomers for research requiring advanced adaptive optics. This altazimuth-mounted, 3.7 m telescope is ultralight (less than 40 metric tons total weight) and

■ *A single, large equatorial mount carries the .25 m coronograph, a spectrometer and a photopolarimeter. A new, .5 m coronograph is also planned for Haleakala observatory.*

equipped with highly efficient adaptive optics. A 100 mm flexible mirror is at the heart of the system. Controlled by more than 900 actuators, it can execute corrections every 50th of a second. With its ultrafast guidance system, the AEOS telescope can track any object in the sky and provide image resolution down to a few centimeters of secret Russian spy satellites and the U.S. space shuttle.

Among other missions, this telescope tracks orbital debris of potential danger to the International Space Station. The facility is also increasingly being used for research on asteroids and the satellites of the outer planets.

Haleakala Observatory was the first facility involved in the Near Earth Asteroid Tracking (NEAT) program. Sponsored by NASA and the U.S. Air Force, the goal of NEAT is to identify and track 90 percent of asteroids whose orbits intersect or have the potential to intersect that of Earth, and which might collide with Earth in the future. This

■ *A 1.2 m telescope at Haleakala Observatory is dedicated to NEAT, a program sponsored by NASA and the U.S. Air Force to identify and track near-Earth asteroids and other objects. A similar facility is operated from Mount Palomar.*

program, which has been running at Haleakala since 2000, uses a 1.2 m telescope equipped with CCD imagers. An identical facility was established at Mount Palomar Observatory in California. It is estimated that about 1,000 near-Earth asteroids larger than 1 km exist, more than 500 of which have been identified through the NEAT program. Once the initial goal is reached, a second much more ambitious program will begin to identify the tens of thousands of smaller but still dangerous asteroids in the 100 m to 1 km size range. This is likely to require a much more powerful telescope however.

None of the asteroids discovered so far are likely to collide with the Earth in the foreseeable future. Nevertheless, given our planet's past history, including geologic evidence for several dozen major impacts, and biological evidence of both ancient and more recent mass extinctions, scientists and politicians alike are certain that near-Earth asteroids pose a real risk to our planet. ■

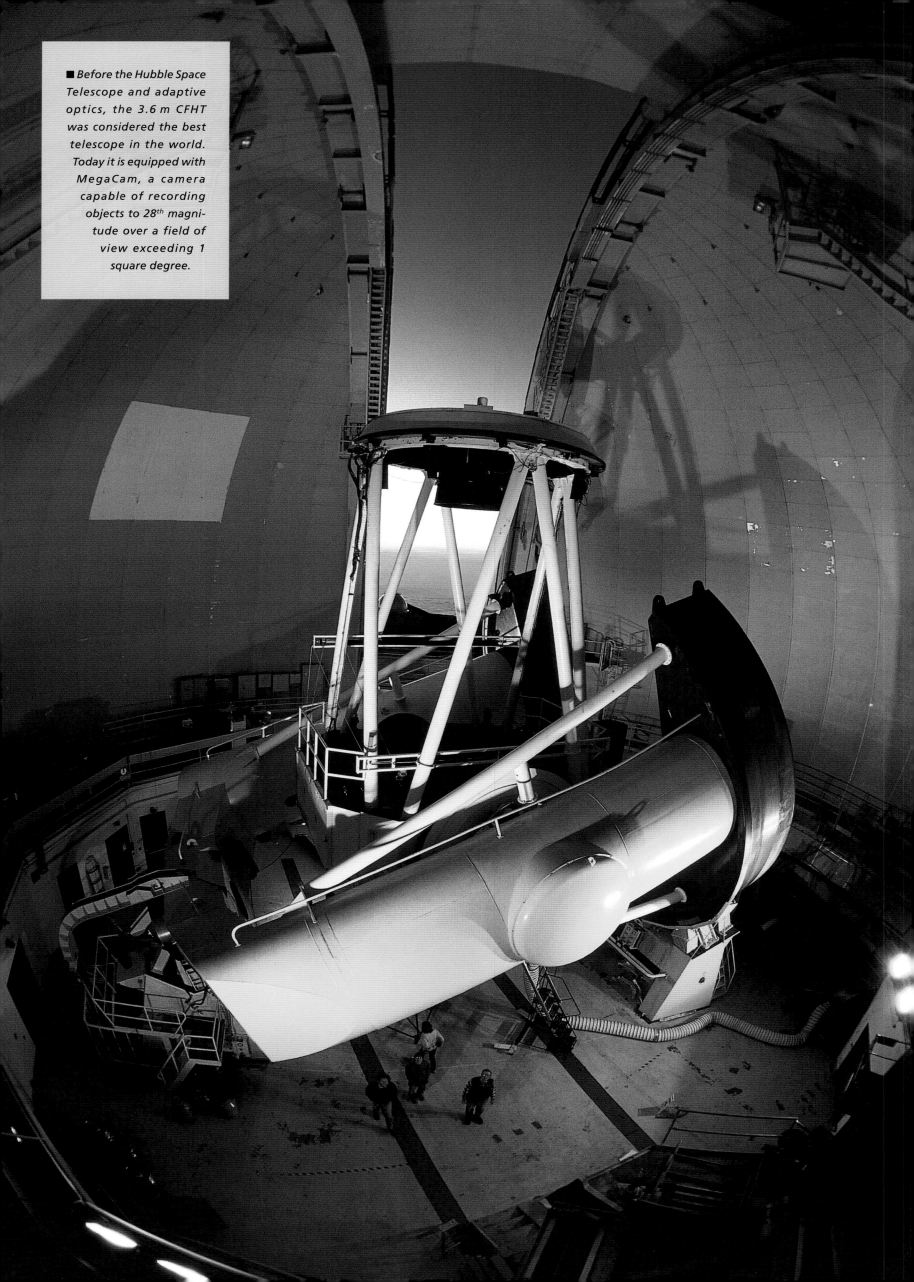

■ *Before the Hubble Space Telescope and adaptive optics, the 3.6 m CFHT was considered the best telescope in the world. Today it is equipped with MegaCam, a camera capable of recording objects to 28th magnitude over a field of view exceeding 1 square degree.*

The Mauna Kea Observatory

After the veil of mist shrouding Hawaii's main island has at last lifted to reveal a panorama of gleaming white beaches, surfers and fishing boats basking in the sunlight, one's eyes are inevitably drawn to the summit of Mauna Kea. There, some 4,200 m above sea level, are the shiny white domes of this famed Hawaiian observatory, the largest astronomical facility in the northern hemisphere.

It is quite a challenge to reach this astronomical haven. To gain access to some of the finest skies on Earth, you must leave the shores of the Pacific and snake your way up Saddle Road, which twists and turns between ranchlands and black lava beds to more than 2,000 m above sea level. This point separates the island's two great volcanoes, Mauna Kea and Mauna Loa. Once there, visitors may be inclined to quit and turn back, since the perpetual cold can easily dampen one's enthusiasm. Mauna Kea, the "White Mountain," is sacred to native Hawaiians and worthy of the name, since there are still 25 km and more than 2,000 m to climb along a rough,

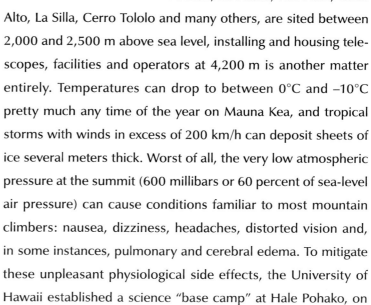

■ *Mauna Kea Observatory is the most important astronomical facility in the world today. At 4,200 m above sea level, the site enjoys more than 300 clear nights a year.*

narrow road before the summit is finally reached. As you drive your 4x4, the only type of vehicle capable of handling the terrain and rarefied air, you soon discover what it is astronomers are seeking here. As you emerge from the mist at about the 3,000 m level, you suddenly enter a very Mars-like scene, a desert of basaltic and ocher-colored cinder. In this totally arid environment there are no signs of birds or insects and not even a hint of vegetation. The sky above is a deep royal-blue color, an extraordinary condition that prevails more than 300 nights per year.

As you look about you some 4,200 m above sea level, short of breath and a little haggard, you are confronted with a surreal, Dali-esque panorama: a vast desert with giant mushrooms seemingly sprouting all around you. They are the observatory domes and

UNITED STATES

radio telescope dishes lining the crests and valleys of this enormous volcanic crater.

In 1964 when the American astronomer Gerard Kuiper and his team discovered this Mauna Kea site, none of the world's great modern observatories existed as yet. Moreover, at a time when an orbiting telescope high above clouds, turbulence and smog was still a dream, no one imagined that such truly calm, pristine skies would ever be found on an island in the Pacific Ocean. Once discovered, Mauna Kea had to be progressively readied for use. First, a sizable roadway had to be constructed, winding its way across difficult terrain all the way to the summit. This had to be done with minimal damage to a site that is sacred to native Hawaiians, who did not want their white mountain desecrated. Mauna Kea was designated as a "scientific preserve" to protect it against development. Finally, astronomers faced a totally new challenge: extreme altitude. Although most of the world's large observatories, including Mount Palomar, Cerro Paranal, Cerro Pachon, La Palma, Kitt Peak, Calar Alto, La Silla, Cerro Tololo and many others, are sited between 2,000 and 2,500 m above sea level, installing and housing telescopes, facilities and operators at 4,200 m is another matter entirely. Temperatures can drop to between 0°C and –10°C pretty much any time of the year on Mauna Kea, and tropical storms with winds in excess of 200 km/h can deposit sheets of ice several meters thick. Worst of all, the very low atmospheric pressure at the summit (600 millibars or 60 percent of sea-level air pressure) can cause conditions familiar to most mountain climbers: nausea, dizziness, headaches, distorted vision and, in some instances, pulmonary and cerebral edema. To mitigate these unpleasant physiological side effects, the University of Hawaii established a science "base camp" at Hale Pohako, on the slopes of Mauna Kea. Located at 2,900 m, this facility has become a comfortable haven for astronomers and technicians on their way to the observatory, and a mandatory stopover point for all long-term visitors to the summit. Workers must spend at least one night or day at Hale Pohako before ascending to the top, long enough to at least partially acclimatize to the rarefied atmosphere. There are no long-term accommodations at the summit for astronomers, who must retrace the winding 15 km path to Hale Pohako as soon as their work is done.

The observatory's beginning was very modest indeed. Ironically, when its first researchers began observations at the summit, the University of Hawaii did not even have an astronomy department. The first telescopes installed at the site in 1968 were two small .6 m reflectors. They were part of an intensive effort to study the Moon and planets in preparation for the first exploratory space probes to various solar-system objects. Astronomers soon came to appreciate the extraordinary potential of Mauna Kea, with almost six months of clear night skies per year and exceptionally steady seeing. In addition, due to its high elevation, the site provides extremely dark skies, ideal for observing faint distant objects like galaxies and quasars. At 4,200 m the atmosphere becomes transparent to infrared and submillimeter wavelengths, something not possible at lower elevation. These windows to the Universe remained largely unexplored before the 1970s, but have since been opened widely, particularly at Mauna Kea.

In 1970, the University of Hawaii inaugurated a 2.2 m optical and infrared telescope, and in 1979 two new infrared-detecting telescopes entered service at the observatory. The first was the NASA 3 m Infrared Telescope Facility (IRTF). The second, the 3.8 m United Kingdom Infrared Telescope (UKIRT), marked the beginning of the internationalization of the observatory. The year 1980 inaugurated a new chapter in the

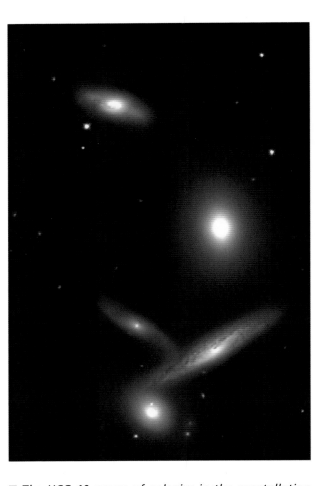

■ *The HCG 40 group of galaxies in the constellation of Hydra is some 300 million light-years from us. This infrared image is a 15-minute exposure taken with the 8.3 m Subaru telescope. In a few hundred million years these interacting galaxies are likely to fuse into one giant elliptical galaxy.*

annals of astronomy, with the establishment of the Canada–France–Hawaii Telescope (CFHT). This 3.6 m instrument, adapted for both visible and infrared wavelengths, was the first to take full advantage of the exceptional seeing and transparency on Mauna Kea. Results did not take long to materialize. Between 1980 and 1990, the CFHT was considered the best telescope in the world for distant galaxies, with resolution often exceeding 0.5 arcsec and magnitude limits of better than 27. Astronomers took full advantage of this situation to observe things hitherto not seen: galactic jets, gravitational lensing effects and fully resolved stars in distant galaxies. Above all, however, the CFHT gave astronomers an opportunity to experiment with new optical technologies. In the 1980s advances in optics and electronics made it possible to explore the sky from Earth with a clarity thought possible only from space. Adaptive optics, which had been developed first by the U.S. military and later at the European Southern Observatory in La Silla, Chile, produced truly spectacular results at Mauna Kea.

In its 20 years of service, the CFHT has undergone several upgrades and rejuvenations. Spectro-imaging devices, such as MOS-SIS, Tiger and OASIS have replaced its original, relatively low-performance spectrographs. In 1996 the CFHT was equipped with a very efficient adaptive optics system called PUEO. By minimizing the effects of bad seeing, this device was ideally suited for monitoring several close binaries at once, or observing stars with unprecedented 0.12 arcsec resolution at the center of our galaxy, close to its giant black hole. Today, superseded by much larger telescopes with greater resolution and sensitivity, particularly in terms of the spectral analysis of extremely faint objects, the CFHT mission has radically changed. It has been converted into a widefield instrument, a sort of super-Schmidt camera. A new CCD camera was installed at its focal

■ *California astronomers have installed two identical 10 m telescopes on Mauna Kea, which will soon be joined into a giant interferometer. The hexagonally segmented mirror of the Keck II telescope is shown here. This telescope weighs 300 metric tons.*

plane in 1999, the CFH12K, with field coverage of 42 x 28 arcmin, a section of the sky considerably larger than the apparent area of the full Moon. This expanded the telescope's capabilities to encompass the Solar System, the Milky Way and the large-scale structure of the Universe. The success of the CFH12K led to development of MegaCam, an even more powerful instrument, installed in 2003. This is the largest astronomical camera ever built. Consisting of a mosaic of 40 individual CCD chips of 4,500 x 2,000 pixels each, the device is .3 m long and wide and holds a total of 360 million pixels.

This huge camera, weighing nearly a metric ton, was installed at the 15 m prime focus of the CFHT, the same focal plane originally used for traditional photography but long since abandoned. Thus some 20 years after its initial installation this venerable telescope has resumed its original mission, widefield imaging, only this time using truly advanced technologies. MegaCam's field coverage exceeds 1° x 1°, equivalent to four full Moons. To ensure uniform image quality across such a wide field, an unusual requirement for a telescope of this size, an .85 m field corrector lens had to be constructed, the largest lens of this type ever ground. The system is also equipped with adaptive optics, with actuators applying corrections 10 times per second to counter atmospheric turbulence. MegaCam is designed to provide angular resolution of at least 0.5 arcsec and reach 28th magnitude with very long exposures.

After the initial phase of telescope construction on Mauna Kea, culminating with the CFHT, a myriad of additional facilities followed.

THE KECK TELESCOPES: A GAMBLE THAT PAID OFF

Initially no one really had much faith in astronomer Jerry Nelson's idea that a truly large telescope mirror could be made by combining several smaller segments into a single array. In fact, apart from the Kecks, only the Spanish Grantecan at La Palma and the twin spectroscopic telescopes, Hobby-Eberly and the Southern African Large Telescope (SALT) have used this design. All other giant telescopes currently in operation have monolithic mirrors around 8 m in diameter. This appears to be the practical size limit most optical designers prefer. However, segmented mirrors like the two Kecks can be made larger and work very well. Each of the 36 hexagonal segments is 1.8 m in diameter, which in combination provide an overall aperture of 10 m. The mosaic is supported by 108 computer-controlled thrusters and 168 actuators, which monitor and align all optical surfaces twice per second. Individual mirror segments are linked by a series of sensors that monitor electrical charge differences between overlapping metal plates and convey any positional variations to the thrusters for immediate correction. The precision attained this way is quite remarkable, since the system maintains surface accuracy to within a few nanometers across the full 10 m aperture.

Wider access to the electromagnetic spectrum was secured in 1987 with the construction of two radio telescopes, the 15 m joint British/Dutch/Canadian James Clerk Maxwell Telescope (JCMT) and the 10.4 m Caltech Submillimeter Observatory (CSO). These instruments extended existing visible-range instrumentation at the observatory into the infrared and submillimeter ranges. In 1993, the westernmost 25 m antenna of the Very Long Baseline Array (VLBA) was installed. This huge radio interferometer, which stretches across the entire United States, has been used to observe the cores of radio galaxies and quasars at centimeter wavelengths. An entirely new submillimeter interferometric array, the Submillimeter Array (SMA), began observations in 2003. Consisting of eight 6 m dishes, this array will serve as a virtual facility with an effective aperture of 600 m. The SMA will generate images in the millimeter and submillimeter wavelength range, between 0.3 and 1.3 mm, at angular resolution from 0.2 to 2 arcsec.

As we begin the 21st century, Mauna Kea is among the most coveted astronomical facilities in the world. This is especially true in the northern hemisphere where, much to the dismay of astronomers, virtually no truly outstanding sites like it remain. In Europe La Palma Observatory in the Canary Islands could still accommodate additional major facilities, and Mount Graham Observatory in Arizona is just coming on line. That leaves only Mauna Kea, which has acquired the bulk of giant telescopes in recent years.

Foremost among these are the twin 10 m Keck telescopes. They were placed into service in 1993

and 1998 respectively, are managed by the University of California and can observe at both visible and infrared wavelengths. Each 10 m primary mirror is segmented, consisting of a mosaic of 36 smaller hexagonal mirrors 1.8 m in diameter. The two telescopes are separated by an 85 m baseline and will operate as a powerful infrared interferometer. Following the lead of the European VLTI project at Cerro Paranal in Chile, which will link four 8 m telescopes into a virtual telescope of 200 m effective aperture, proposals have been submitted to add four to six smaller telescopes near the existing twin telescopes to expand the Keck optical interferometer. Similar to the European effort, this facility should be completed by 2005 or 2006. An undertaking of this nature is technically very difficult. To minimize atmospheric seeing effects, each 10 m telescope will be equipped with adaptive optics coupled with laser-projected artificial stars. Keck I is already using such a system. When all systems are fully operational, light from all five telescopes in the array will be combined into a virtual instrument of several hundred meters aperture. It is expected that this will provide resolution in the order of 0.001 arcsec at a wavelength of 1 μm and 0.01 arcsec at 10 μm.

Sited near the two Keck domes on the same side of the volcano's crest is Subaru, the great 8.3 m Japanese national telescope, also among the most important astronomical instruments of the early 21st century. Unlike the two Keck facilities, Japanese astronomers opted for the design pioneered by the European Southern Observatory by using a lightweight "real" mirror, 8.3 m in diameter but only .2 m thick. Subaru became operational in 2000 and represents a giant leap forward for Japanese astronomers, whose largest telescope up to that time was a 1.8 m reflector.

The year 2000 also saw the inauguration of another huge telescope on Mauna Kea: Gemini North. In order to free enough space for this new facility on the increasingly crowded mountaintop, the now antiquated .6 m NASA telescope had to be removed. Gemini North is the first of a pair of such telescopes built by an international consortium. Its twin, Gemini South, is located in the Chilean Andes to cover the southern hemisphere. Like the four components of the European VLT, Gemini North has an 8.2 m mirror, ground and polished by the French company Reosc. During the inauguration of the Gemini telescope, astronomers decided that its "first light" test would be undertaken with full adaptive optics in operation. The results were spectacular, as the instrument attained an average of 0.1 arcsec resolution at a wavelength of 2 μm, nearly as good as the Hubble Space Telescope. To get a fuller sense of what this means, 0.1 arcsec is equivalent

to resolving objects only 180 m in diameter on the surface of the Moon. The total project cost, amounting to some $150 million, was financed by the United States, Great Britain, Canada, Chile, Australia, Brazil and Argentina.

It is hard to predict at this point how Mauna Kea Observatory will develop in the long run. Despite appearances to the contrary, space on the summit is limited. Although the mountain's surface extends over several square kilometers, the preferred telescope placement is in a northeasterly direction, in line with the prevailing wind pattern and hence with minimal atmospheric turbulence. This limits available space to two or three cinder cones along the line already well occupied by existing facilities. As indicated before, the very top of the mountain, a stunning 50 m high cinder cone

is off limits; the only other options are to remove smaller existing telescopes in favor of new ones. This already happened with the .6 m telescope for Gemini North, and might well include the 2.2 m University of Hawaii telescope, the 3 m IRTF, the 3.6 m CFHT and the 3.8 m UKIRT in the future, as giant 25 and 50 m telescopes are contemplated. The Institute of Astronomy at the University of Hawaii is naturally eager to continue managing the site and see the construction of instruments considerably larger than their own 40-year-old facility. After all, whenever a new facility comes on line, be it Californian, French, British or Japanese, Hawaiian astronomers are allocated 15 percent telescope time on each!

the 2.2 m University of Hawaii telescope, the 3.8 m UKIRT, and lastly, the .6 m University of Hawaii telescope. In the foreground below are the SMA under construction, the 15 m JCMT submillimeter facility and the 10.4 m CSO.

This assures the 25-year-old institute continued access to the best and most modern astronomical facilities available.

Several multinational efforts are underway to build the next generation of giant telescopes on Mauna Kea in the coming decades. For example, Canada and France are planning the Next Generation Canada–France–Hawaii Telescope (NGCFHT) to replace the aging CFHT. With a 25 m segmented mirror and adaptive optics, this new telescope is expected to attain 0.01 arcsec resolution at 1 μm, and reach objects as faint as 32nd to 33rd magnitude.

Other future giant telescopes planned for either Mauna Kea or the Atacama Desert in Chile include the Giant Segmented Mirror Telescope (GSMT), a 30 m instrument planned by the U.S. National Optical Astronomical Observatory (NOAO), and similar efforts for a California Extremely Large Telescope (CELT) involving the University of California and the California Institute of Technology (Caltech).

Undoubtedly island authorities will continue their efforts to protect Mauna Kea in the coming decades against excessive light pollution. Unlike Oahu with Honolulu and its many beaches, Hawaii's big island is still free of extensive industrial pollution and massive tourism. With its exceptional skies, two 10 m telescopes, two 8 m instruments, three 4 m facilities and an array of radio telescopes, Mauna Kea is without question the largest center of observational astronomy in the world today. ∎

The Cerro Tololo Observatory

A flight of condors circles high above the mountain, casting shadows as big as airplanes onto the gleaming domes of the observatory. Cerro Tololo is truly one of the most spectacular astronomical sites in the world. Its seven domes, ranging in height from 10 to 40 m and bright as snow, are clustered on a gravel pavement covering a half-hectare at most. A first-time visitor, buffeted by high-altitude winds, is hard pressed to describe the scene; is it a heliport, a Manhattan highrise or the platform of an alien spacecraft?

The Chilean Andes have become a haven for modern astronomy. No less than five large observatories have been built here in the past 40 years, directly between the Pacific Ocean and the high volcanic cordillera to the east. It was probably American astronomers who first realized the exceptional potential of this 2,000 km long stretch of the Andes. The first astronomer to observe from Chile, James Gillis, spent five months near Santiago in 1849, describing the celestial treasures of the southern hemisphere. Then in 1903 a .9 m telescope, very large for its time, was installed on a hilltop in Santiago. This Lick Observatory instrument was shipped from San Francisco to the Chilean port of Valparaiso, and then hauled piecemeal to the capital by mule train.

In 1950 astronomers Federico Ruttland and Hugo Moreno began the first systematic evaluations of seeing and meteorological conditions at potential observing sites in Chile. They represented a consortium of American universities interested in establishing a southern hemisphere counterpart to Kitt Peak Observatory in the north. In 1962 Cerro Tololo, located at 30° south latitude and about 50 km from the Pacific Ocean and the town of La Serena,

CHILE

■ *Cerro Tololo was the first large observatory to be established in Chile. Its main instrument, the 4 m Blanco telescope, is the twin of the Mayall reflector at Kitt Peak.*

was selected as the site for the Cerro Tololo Interamerican Observatory (CTIO).

At an elevation of 2,000 m, this observatory was financed and built by the United States but has hosted astronomers from all across the Americas, including Canada, Chile and occasionally Brazil, Venezuela, Mexico, Argentina and other countries. Top researchers have returned repeatedly to the 40-year-old observatory. The largest instrument was installed at the site in 1976: the Blanco telescope, a twin of the 4 m Mayall telescope at Kitt Peak. Complete with its classic equatorial mount, this telescope's total movable mass is close to 340 metric tons. Like most instruments of that generation, it is used primarily today for widefield work and long-term survey operations of selected areas of the sky. To accomplish this, a field flattener has been installed at the telescope's f/2.87 prime focus and an 8,000 x 8,000 pixel CCD camera. The field of view of this combination exceeds 30 arcmin, and resolution is often better than 0.7 arcsec under optimal conditions. This camera has been particularly useful in detecting extremely faint asteroids and distant supernovae. Supernovae with well-defined brightness and luminosity characteristics are used as standard candles by astronomers to help refine such parameters as the Hubble constant (H) and the cosmic density ratio (abbreviated by Greek letter omega, Ω).

The Blanco telescope supports a special 50,000 resolution spectrograph, Hydra, at its f/8, 32 m Cassegrain focus. Sensitive to wavelengths from 330 to 1,100 nm, this instrument is equipped with 138 optical fibers that can be repositioned in less than half an hour to any object in the spectrograph's 40 arcmin field of view.

In 2001 an Anglo-American team from Berkeley, Edinburgh

and Cambridge universities made a truly spectacular discovery with this instrument. The researchers were looking for low luminosity stars in the halo of our galaxy, stars that could collectively amount to between 300 billion and 400 billion solar masses. Although such stars had eluded detection for nearly 70 years, their presence had long been surmised through gravitational disturbances in the outer fringes of the Milky Way. Scientists began the search by analyzing photographic plates from a 3,000-square-degree Schmidt-telescope sky survey, carried out over the preceding 30 years at Siding Spring Observatory in Australia. This was followed by a spectroscopic survey with the Blanco telescope of several hundred stars thought to be part of the galactic halo because of their high apparent motion. Some 40 white dwarf stars were discovered this way, stars that are less than 500 light-years from us. This indicated that the full galactic halo might contain anywhere from 5 billion to 100 billion such stars, which account for 3 to 35 percent of its total mass. If the latter value were confirmed, the mystery of the "invisible" stars in the galactic halo would be solved. Physicists have estimated that the remaining dark matter in the cosmos consists of a combination of exotic particles that do not interact with ordinary matter, including neutrinos, neutralinos, photinos and others.

With the exception of the 4 m telescope, most other instruments at Cerro Tololo are rather small and include .6, .9, 1, 1.3 and 1.5 m instruments. The 1.3 m and its twin on Mount Hopkins, Arizona, participated in the Two Micron All Sky Survey (2MASS). Finished in 2001, this survey helped map all infrared sources in the sky, at wavelengths of 1.25, 1.65 and 2.16 μm. The two telescopes recorded more than 300 million stars and galaxies to 15th magnitude. The resulting 2MASS data, 24 terabytes in all, are currently being analyzed at the University of Massachusetts and the Infrared Processing and Analysis Center (IPAC) in California.

Among other things, this large infrared database will provide more information on red and brown dwarf populations and distribution in the central regions of our galaxy. The 2MASS project has already led to the discovery of several brown dwarfs, as well as distant galaxies and quasars hitherto masked by thick clouds of interstellar dust opaque to infrared radiation. This survey, sponsored by National Science Foundation and NASA, will lay the foundation for future generations of space-based telescopes like Herschel, SOFIA, SIRTF (Space Infrared Telescope Facility; since renamed the Spitzer Space Telescope), and of course the NGST (Next Generation Space Telescope), all intended to observe the cosmos at infrared wavelengths. ■

> Adaptive optics

When Galileo turned his telescope skyward in 1610 and first saw Jupiter's four large moons he also discovered the first celestial objects not visible to the naked eye. This opened an entirely new window to the cosmos and put astronomers on a steady path to develop bigger and better telescopes to see ever fainter and more distant objects in the Universe.

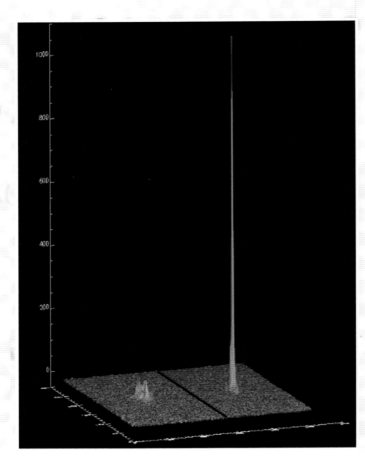

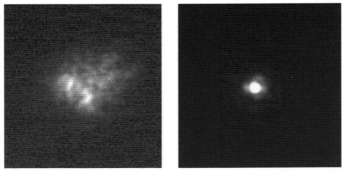

● This illustration shows the same star imaged with and without adaptive optics. The star appears 0.5 arcsec in diameter on the left, and 0.07 arcsec on the right when the Nasmyth Adaptive Optics System (NAOS) is applied.

In general, atmospheric conditions aside, the larger the diameter or aperture of a telescope the higher its resolving power or ability to discern fine detail. Atmospheric interference can be overcome by launching telescopes into space, but this is a costly alternative and obviously limited by the payload capacity of the launch vehicle. Fortunately, the development of adaptive optics over the past few decades has been so successful at compensating for atmospheric turbulence that all modern telescopes are now equipped with this technology.

WHEN EVEN THE LARGEST TELESCOPES SEE ONLY FUZZY IMAGES

Every amateur astronomer is only too familiar with really bad seeing. On nights like that, images in a .6 m telescope are no crisper than those delivered by .2 or .3 m instruments. Unfortunately this also applies to the largest ground-based instruments, even though the laws of optics tell us that resolution is directly proportional to the diameter of the telescope. The problem is our atmosphere. As light rays from celestial objects enter our atmosphere, they are dispersed and diffracted to varying degrees by turbulent air cells. Although minute, these distortions accumulate as the light passes through many kilometers of air, adding up to several micrometers by the time the light rays reach the telescope. This can be enough to significantly degrade both image quality and resolution.

In the 1950s, physicist Harold Delos Babcock proposed a simple solution to this problem by suggesting that reversing the dispersive effects of atmospheric turbulence could restore image quality. Although he had outlined adaptive

● *This spectacular image of the Sun's surface was obtained with the Dunn Solar Telescope (DST) at Sacramento Peak Observatory. The .76 m telescope is equipped with adaptive optics involving some 100 actuators. This image taken at 500 nm and 0.15 arcsec resolution shows detail approximately 110 km in diameter on the surface of our star.*

optics in principle, putting this into practice took a very long time. Such technology would have to simultaneously analyze and correct light distortions, and do this under constantly changing atmospheric conditions. This took many years to perfect and was initially restricted to secret military applications. Consequently adaptive optics did not become widely available for astronomical applications until the late 1980s.

THE BASICS OF AN ADAPTIVE OPTICS SYSTEM

The wave front of a ray of light is perpendicular to its direction of travel. Adaptive optics are designed to correct wave front distortions as light rays travel through the atmosphere. This is accomplished by continuously monitoring distortions relative to an ideal light wave and then computer-correcting for this several times per second.

Such systems require three key components: a wave analyzer, a real-time analysis and corrective optics. To compensate for often rapid wave-front fluctuations, all components are computer-controlled and typically execute several corrections per second. Image quality ultimately depends on the rapidity with which the adaptive system responds. With a modern 8 or 10 m class telescope, several million corrections per second may be required. This was simply not possible in the 1950s when Babcock lay the foundations of adaptive optics.

● Saturn as imaged by the 8.2 m Yepun telescope, equipped with the NAOS adaptive optics system and the CONICA camera. The image is a composite of exposures taken with 1.6 µm and 2.2 µm filters. This is the highest-resolution image (0.07 arcsec) ever taken of the ringed planet by an Earth-based telescope.

HIGH-PERFORMANCE ADAPTIVE OPTICS FOR INFRARED WORK

Adaptive optics technologies today are particularly well suited for imaging in the infrared. Since atmospheric convection cells are quite small, heavy turbulence impacts the shorter visible wavelengths more noticeably. Because of this, adaptive optics have to operate more rapidly and effectively with visible light than at infrared wavelengths to provide comparable image quality. For instance, an 8 m telescope operating in the visible range at 0.6 µm requires 6,400 actuators for effective image correction, but only 250 actuators at 2 µm in the infrared.

USABLE FIELD WITH ADAPTIVE OPTICS

The first step when using adaptive optics is to measure the amount of wave-front distortion in light rays from a "reference star," typically either a medium-bright star or slightly extended objects like planetary satellites or the nuclei of distant galaxies. Once this is determined, a corresponding amount of image

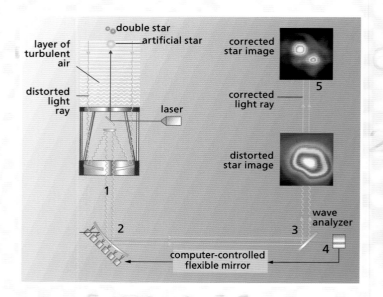

● The diagram at left summarizes how adaptive optics work. First, the incoming light beam (1) is focused onto a flexible mirror (2) and from there onto a beam splitter (3). Part of the light goes to a computerized wave analyzer (4) and the rest is used for observational data. Once the amount of wave-front distortion is assessed relative to an ideal wave, computers compensate for this through rapid deformations of the flexible mirror, which then sends the corrected light rays to the telescope's recording camera or spectrometer (5). The photo on the right shows the NAOS adaptive optics assembly of the 8.2 m Yepun telescope array.

correction can be applied. Since the resulting mages will generally appear sharpest near the reference star and fall off in adjacent areas influenced by neighboring convection cells, it is best to use the target itself as reference object whenever possible. In situations where this cannot be done, as with faint or very diffuse objects, then adjacent stars must be used, preferably those within 30 arcsec of the target and likely to lie in the same air column. Adaptive optics using laser-projected reference "stars" are increasingly favored since they can be pointed at any target and do not require natural reference stars. Though more complex to use in practice, laser-guided adaptive optics are likely to see more and more use in future.

ADAPTIVE OPTICS COME INTO THEIR OWN

The first adaptive optics system used specifically for astronomical purposes was developed in France for the 3.6 m telescope of the European Southern Observatory. The first such systems, Come-On and Adonis, used 19 actuators and 52 actuators respectively. Shortly after that, a system called PUEO (Hawaiian for owl) was successfully adapted to the 3.6 m CFHT at Mauna Kea. The fourth telescope in the VLT array is equipped with the 185-actuator system NAOS,

like the system installed on the large Keck telescope. All the new large telescopes, including Gemini, Subaru and the Large Binocular Telescope, are similarly equipped to ensure that these instruments will deliver the best images possible.

Simulation studies with next-generation telescopes like MAXAT, CELT and OWL (all giant 50 to 100 m class instruments) will not be able to function *without* adaptive optics. Since these systems are expected to perform better than existing designs, entirely new technologies will have to be developed.

FUTURE DEVELOPMENTS

A number of advanced adaptive optics designs are under development, not only for widefield applications but also for improved performance in visible light. Systems like these will be multifaceted and capable of measuring a number of wave fronts simultaneously, using either natural or artificial reference stars in the same field as the object under study. This will make it possible to apply image correction across fields several arc minutes wide, rather than just a few arc seconds, as is currently the case. This type of advanced technology will be essential for future-generation telescopes and support instrumentation. ■

● The resolving power of modern Earth-based telescopes has reached the point where meteorological and geological processes on distant planets can be imaged. The figure at left shows Jupiter's moon Io as imaged by the Keck II telescope with adaptive optics. The figure at right shows Io imaged by the Galileo spacecraft in 1995 during its visit to the Jovian system. The Keck telescope image, taken at a wavelength of 2.2 μm, attained better than 0.06 arcsec resolution.

The Cerro Pachon Observatory

CHILE

The impressive facade of Cerro Pachon overlooks the Rio Elqui, which irrigates all villages in the valley including Serena, where it empties into the Pacific Ocean. Reaching nearly 2,800 m above sea level, Cerro Pachon is part of the frontal range of the Andes some 50 km to the east, where many peaks rise well above 5,000 m. From this spectacular vantage point, the tiny white domes of Cerro Tololo Observatory can just be made out some 600 m below and 10 km farther west.

The long, high desert crest of Cerro Pachon was selected as the ideal location for Gemini South, one of the most powerful modern telescopes in the world. Project Gemini began in the early 1990s as a collaboration between national astronomical institutes in the United States, the United Kingdom, Canada, Chile, Australia, Brazil and Argentina. The goal was to establish a powerful modern astronomical facility in the southern hemisphere. Under the umbrella of the Association of Universities for Research in Astronomy (AURA), the seven participating nations have

■ *This is a view inside the dome of the Gemini North telescope, which entered service in 2002. Situated 2,715 m above sea level on Cerro Pachon, this facility lies just above the observatory on Cerro Tololo.*

committed $200 million, with the United States contributing half the funds, to build two identical 8.1 m telescopes. Gemini North saw first light atop Mauna Kea in Hawaii in 2000, and Gemini South at Cerro Pachon in 2002.

The brand new Cerro Pachon Observatory is truly impressive. Set near the edge of the mountain, the dome is 36 m in diameter and 46 m high. The total weight of the telescope including its altazimuth mount is 340 metric tons, nearly the same as the considerably "smaller" 4 m telescope at Cerro Tololo. The telescope mirror, like the one for Gemini North, was ground

and polished by the French firm of Reosc, which also polished the four mirrors for the European VLT. The 8.1 m mirror is a glass meniscus only .2 m thick and weighs only 22 metric tons. Its surface is accurate to within 15 nm. In addition to its Cassegrain focus, Gemini South is equipped with two f/16, 128 m Nasmyth foci and extensive support instrumentation. The telescope's pointing accuracy is better than 3 arcsec and its tracking accuracy near 0.01 arcsec.

The Gemini telescopes were designed primarily for infrared work. Sited at 4,200 and 2,800 m respectively, both instruments enjoy exceptionally dry, transparent skies. Like Cerro Tololo, Cerro Pachon is an uninhabited ecological preserve covering some 350 square kilometers and is essentially free of air and light pollution: in short, ideal for astronomy.

In 2002 construction began on another dome not far from Gemini South. This will house the Southern Observatory for Astrophysical Research (SOAR), a joint United States–Brazilian facility. Slated for completion in 2005, this 4.2 m altazimuth-mounted telescope will support and complement Gemini South at optical wavelengths. Since the two telescopes have identical f/16 focal ratios, support equipment will be readily interchangeable between them.

Researchers at Gemini South are presently concentrating on infrared observations, since they are fully equipped for that. In addition to this, the telescope's Cassegrain focus supports HRSO, a 50,000 resolution spectrograph that spans a wide range of wavelengths from 300 nm to 1 μm, from ultraviolet through visible and near-infrared portions of the spectrum.

■ The multinational project Gemini involves twin 8.1 m telescopes, one on Mauna Kea in the northern hemisphere, the other on Cerro Pachon in the southern hemisphere.

Phoenix, a spectrograph first used at Kitt Peak, is an instrument with 50,000 to 100,000 resolution, intended primarily for studies of bright stars at wavelengths of 1 to 5 μm. A second infrared spectrograph, OSCIR, previously used on the Keck II telescope in Hawaii, provides resolution of 100 to 1,000 between 8 and 25 μm and will study extremely faint stars and galaxies at Cerro Pachon. Flamengos, another infrared spectrograph, can analyze 50 stars simultaneously with resolution around 1,000 in the 1 to 2.5 μm range.

In 2003 Gemini South was equipped with T-ReCS, a very high-performance spectrograph-imager specifically designed for this telescope. In addition to imaging in the infrared between 8 and 26 μm, this instrument can take 100 to 1,000 resolution spectra of very faint objects. T-ReCS should produce exceptionally high-resolution images for a number of reasons. First, infrared wavelengths are not as affected as visible light by atmospheric turbulence. Second, Gemini South is equipped with a novel *tilt-flip* secondary mirror system for rapid image correction, and third, the telescope's tracking accuracy is extremely high. As a result, the telescope is expected to attain 0.3 arcsec resolution at 10 μm, close to its theoretical performance level.

As has been successfully demonstrated with Gemini North, this new adaptive optics system is also very effective for near-infrared work. Working at f/33 and an impressive 268 m focal length, the system has a fully corrected field of view of nearly 80 arcsec when operating between 0.85 and 2.5 μm. Both Gemini telescopes have attained 0.7 arcsec resolution at 2.2 μm, making them ideal for studies of star formation and protoplanetary systems. They will also penetrate deep into the galactic center into areas obscure at visible wavelengths, but clearly very complex and dynamic when observed at infrared and radio wavelengths.

An entire new area of astronomical research is now open to the Gemini telescopes and other 8 to 10 m class telescopes equipped with adaptive optics, like Subaru, the two Kecks and the VLT array at Cerro Paranal. No other Earth- or space-based instrument can come close to these giants working in the infrared with adaptive optics. Even the Hubble Space Telescope's NICMOS camera operating at 2.2 μm attains only 0.2 arcsec resolution, while Gemini can reach 0.07 arcsec.

In the coming years AURA hopes to develop a new-generation, multilinked system of adaptive optics for Gemini South. This would bring its fully corrected field of view close to 2 arcmin, making this ground-based instrument almost a "space" telescope. Such technology might well raise Gemini South's capabilities in the next 15 years to rival Hubble and even the Next Generation Space Telescope (NGST). Even more important though, these developments will lay the foundation for entirely new adaptive optics systems for giant telescopes of the future. ■

Las Campanas Observatory

CHILE

From our 2,300 m high vantage point at Las Campanas, the neighboring European Southern Observatory (ESO), a scant 30 km south of us, is indeed an impressive sight. All of its domes are visible from here, as if parading along the crest of La Silla. Compared to the technical and financial resources that Europeans have invested in the ESO, Las Campanas Observatory is both smaller and more modestly supported. For many years Las Campanas felt a little like Tom Thumb compared to La Silla and the more southerly Cerro Tololo Observatory. All that changed, however, at the turn of the millennium when the two new Magellan telescopes were built at Las Campanas.

The observatory was established in the late 1960s by the Washington-based Carnegie Institution, essentially as a southern hemisphere complement to Mount Wilson Observatory in California. Las Campanas is a high desert peak with exceptionally calm and clear atmospheric conditions and covered with sun-baked rocks shielding tarantulas as big as a person's hand. The facility began modestly enough. The first two instruments, installed in 1971,

■ *The two new 6.5 m telescopes at Las Campanas are shown above. These very compact but powerful instruments only weigh 150 metric tons each.*

were a domed 1 m telescope and a simple .3 m astrocamera under a sliding roof observatory. The superb Irenee du Pont telescope was installed nearby in 1977. This classically mounted 2.5 m telescope had only a single f/7.5 Cassegrain focus of 19 m focal length. Before the advent of CCD cameras, it used a special field flattener and .3 x .3 m photographic plates.

The du Pont telescope is best known today for obtaining the very first images of another solar system some 63 light-years from the Sun. In 1983 the infrared sensitive satellite IRAS discovered several stars radiating very strongly at these wavelengths. Las Campanas astronomers Brad Smith and Richard Terrile decided

to look at one of these stars, Beta Pictoris, in more detail with the 2.5 m telescope and an infrared camera. They used a special occulting bar to blot out light from the star itself and discovered a tenuous disk of gas and dust around it. The extended disk probably consists of comets, dust and asteroids, and even preplanetary material, as subsequently revealed by astronomers at the La Silla and Siding Spring observatories.

A few years later an even more dramatic finding garnered Las Campanas Observatory worldwide attention. On February 24, 1987, Canadian student Ian Shelton was photographing the Large Magellanic Cloud. This is a small satellite galaxy, which under really dark skies looks very much like a cloud detaching itself from the Milky Way. He was using the site's smallest telescope for this work, a .3 m instrument readily surpassed by many amateur facilities today. After developing his photograph, the young researcher noted a prominent dark spot on the negative, smack in the middle of the little galaxy, near the famed Tarantula Nebula. Convinced that this was either an artifact or a meteor trail, Shelton stepped outside again to take another look at the Large Magellanic Cloud…sure enough, there it was, a bright star shining steadily amid the ghostly little galaxy. Shelton was not the first to see the supernova, however. Oscar Duhalde, a Chilean technician at the observatory and a seasoned observer, had seen it earlier that night but told no one about it. Credit for the discovery, therefore, correctly went to the Canadian astronomer who alerted the world about it the next day via a telex from Serena to the International Astronomical Union.

Astronomers had been waiting for such an extremely rare event for nearly four centuries. The last recorded supernova in our

■ *This widefield image of M823 in Hydra was taken with the 2.5 m Irenee du Pont telescope. A close up, six-minute exposure of one of the spiral arms of this galaxy is shown at right, taken with the 6.5 m Baade telescope. This 0.4 arcsec resolution image clearly shows several blue supergiants and stars down to 25th magnitude.*

vicinity was a naked eye discovery in 1604 by the famed Johannes Kepler, or well before Galileo even used the first telescope!

After examining images of the Large Magellanic Cloud taken before February 24, astronomers quickly identified Sanduleak −69202 as the exploded star. A blue supergiant, Sanduleak was about 60 million km in diameter; 20 times more massive and nearly a hundred thousand times more luminous than the Sun. Supernova 1987A has been carefully monitored ever since. Over time it declined in brightness, emitted shock waves into the surrounding interstellar medium, and ultimately formed a beautiful expanding gas shell. This object was studied with literally hundreds of telescopes of every kind, both Earth- and space-based, and has accrued more than a million hours of observing time. Even as the supernova was beginning to fade, Carnegie Institution scientists began plans for much larger telescopes at Las Campanas.

THE MAGELLAN PROJECT

The Magellan Project began in 1992, when the Carnegie Institution, in association with the universities of Arizona, Harvard and Michigan, and the Massachusetts Institute of Technology (MIT), decided to build a pair of 6.5 m telescopes at Las Campanas. The two borosilicate mirrors were ground and polished at the University of Arizona's Mirror Lab. Like the 6.5 m mirror of the Mount Hopkins telescope, these were also poured and spun-cast

by astronomer-optician Roger Angel and his team. Both mirrors are lightweight, honeycomb-backed and have ultrafast f/1.25 focal ratios. The two altazimuth-mounted telescopes weigh only 150 metric tons each and are housed in large, 60 m wide domes atop Cerro Manqui, somewhat higher than the original facilities at the observatory.

The Baade telescope saw first light in 2001. Provided with two f/11 Nasmyth foci, this telescope is supported by a spectrograph called IMACS and a MagIC camera. This rapid and easily used camera is intended for observations requiring quick responses, like the sudden appearance of novae or supernovae. MagIC is furnished with a variety of narrowband pass filters centered on 350, 477, 623, 762 and 913 nm, similar to those used by the 2.4 m Sloan Digital Sky Survey telescope at Apache Point Observatory.

The 6.5 m Clay telescope, completed in 2003 and identical to the above, has two 71 m Nasmyth foci and a 98 m Cassegrain focus. With these two large telescopes, the 400 scientists of the multi-university consortium will embark on stellar population studies in nearby southern hemisphere galaxies, including NGC 5128 in Centaurus, NGC 1365 in Fornax, NGC 253 in Sculptor and M83 in Hydra. Another long-term project will be to study galaxy evolution over time. The twin telescopes will catalog spectra of galaxies than formed just a few million years after the big bang as well as much younger galaxies. ■

La Silla Observatory

While heading north along the Pan-American Highway toward Vallenar and Copiago, some three hours out of La Serena, Chilean truck drivers reach La Frontera, the front range of the Andes. The domes of La Silla Observatory lie nearly 2,400 m above them. From this vantage point more than 30 km away, the observatory presents an ever-changing panorama. Just before sunset, the dozen tiny domes appear orange-red before fading into darkness, while during nights of full moon they look like ghostly monoliths along the mountain's ridge. Apart from Hawaii's Mauna Kea, no other location in the world boasts as many telescopes.

As they wind their way from the observatory's private airstrip toward whichever telescope they will use that night, visiting scientists can't help but feel lost in some futuristic mountaintop village. They first pass several buildings that include garages, repair shops, a gas station, a first-aid center, sports facilities and a small factory for liquid nitrogen and helium. The final ascent brings them past an imposing administrative building, optical and electronics labs, a hotel-restaurant and, finally, a single-story dormitory for observers and technicians. The observatory can accommodate about 200 people in total.

For many years before a new observatory was built on Cerro Paranal, La Silla was the flagship facility of the European Southern Observatory (ESO). Despite its long and often laborious development, science has always enjoyed a privileged position among European nations. For example, CERN (the European Center for Nuclear Research) has been the world's leading center for particle physics for half a century. The European Space Agency (ESA) is second only to NASA in space exploration, and now ESO is widely

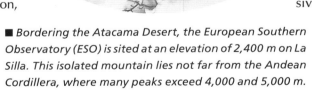

■ Bordering the Atacama Desert, the European Southern Observatory (ESO) is sited at an elevation of 2,400 m on La Silla. This isolated mountain lies not far from the Andean Cordillera, where many peaks exceed 4,000 and 5,000 m.

recognized as the world's largest center for astronomical research. La Silla Observatory and the ESO share a great deal of history. ESO was established in 1962 to provide European astronomers the tools for world-class research. Prior to this, the many antiquated astronomical facilities across Europe simply could not compete with the superior American observatories. The fledgling ESO (still headquartered at Garching near Munich) opted to locate its new facilities far from Europe in the southern hemisphere, a decision based solely on scientific considerations. First, the southern skies feature many of the most interesting astronomical objects, including the Magellanic Clouds and the center of our galaxy. Second, in the 1960s there were few major observatories in the southern hemisphere and the northern skies had already been widely explored by the large American telescopes. After an extensive survey of several arid locations in both South Africa and Chile, La Silla was selected, thanks in part to the success of the U.S. site at nearby Cerro Tololo Observatory.

A HOST OF TELESCOPES

The first group of rather modest-size telescopes became operational at La Silla in 1969 on the side of the mountain bordering the Atacama Desert. A host of additional instruments quickly followed, including two ESO telescopes and several national facilities. La Silla was considered one of the best astronomical sites in the world before better locations were discovered farther north and closer to the center of the Atacama. Far removed from the city of La Serena, more than 160 km to the south, and the village of Vallenar, some 150 km to the north, La Silla enjoys almost 300

■ *The 15 domes of the European Southern Observatory follow the winding crest of La Silla. Nearby is a small town with labs, warehouses, a hotel-restaurant, health center, administration buildings and an airstrip.*

■ *The 3.6 m telescope at La Silla Observatory became operational in 1976. Equatorially mounted, this instrument weighs more than 200 metric tons. This telescope was equipped with Adonis, the first operational system of adaptive optics.*

clear nights a year. To help buffer the mountain against other encroachment, the ESO also purchased an 825-square-kilometer area of land around the observatory.

By the early 1970s the observatory consisted of eight small domes housing a .4 m refractor with an objective prism, two .5 m telescopes, and .6, .9, 1 and 1.5 m telescopes. The first truly international facility was put into service in 1971, a large-aperture Schmidt telescope. This instrument, along with its counterparts at Siding Spring Observatory in Australia and Mount Palomar in California, participated in the first all-sky photographic survey, the Sky Atlas. With its 1 m corrector plate, 1.6 m spherical primary, and 3.05 m focal length, the ESO Schmidt is identical to the famed Palomar Schmidt used for the northern sky survey. With its 5.4 x 5.4 degree field coverage and .3 x .3 m photographic plates, this instrument still needed more than 600 plates to cover the entire southern sky to about 21.5th magnitude. The resulting Sky Atlas represents one of the most significant astronomical undertakings in history. Now fully digitized, the atlas will continue to be an invaluable resource for many decades and serve as a reference "snapshot" taken at the end of the 20th century for future studies of stellar evolution and proper motion in the Milky Way.

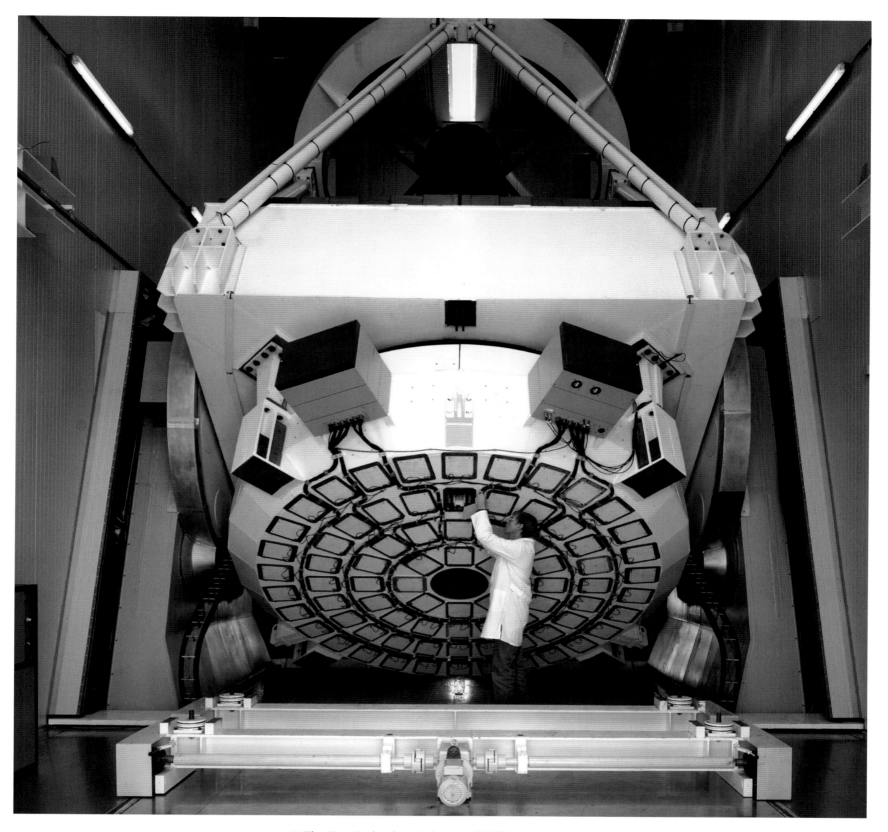

The largest, most powerful telescope at La Silla Observatory was installed in 1976 on the mountain's summit at an elevation of 2,400 m. At that time, this 3.6 m telescope was the fifth largest in the world. Thanks to this remarkable facility, European astronomers were finally in a position to undertake research in astrophysics and cosmology on par with their American colleagues. The ESO, which now includes 11 countries including Belgium, Denmark, France, Germany, Italy, Holland, Portugal, Sweden, Switzerland, Chile and the United Kingdom, has more than a thousand requests annually for telescope observing time.

Because of such demand, the observatory grew even more in the 1980s. A 1.4 m CAT telescope was added, a 2.2 m instrument, a 15 m millimeter wavelength radio telescope (the Swedish–ESO Submillimetre Telescope, SEST) and, finally, another 3.6 m instrument, the New Technology Telescope (NTT).

The NTT saw first light in 1990. Its highly innovative design includes a thin (only .24 m thick), very fast f/2.2 mirror, with two f/11 Nasmyth foci. The telescope is supported by an altazimuth mount and weighs only 120 metric tons, in contrast to 200 metric tons for similar-size equatorially mounted instruments.

While active in research, La Silla also served as a testing ground for the planned ESO Very Large Telescope (VLT) throughout the 1990s, particularly the NTT. As the first instrument equipped with adaptive optics and fully computer-controlled, this telescope was equipped with 100 electronic actuators that adjusted the shapes of its two mirrors every second, thereby ensuring optimal alignment and image quality.

FROM THE NTT TO THE VLT

When first installed, the 3.6 m telescope at La Silla was also equipped with Adonis, the first adaptive optics system in the world. Using some 500 corrections per second to counteract the effects of atmospheric turbulence, Adonis improved image resolution in the infrared to nearly 0.1 arcsec, and provided views of protoplanetary disks around nearby stars comparable to the Hubble Space Telescope. Had it not been for these technological breakthroughs it is unlikely that the European nations would have committed to the VLT project.

When the four 8.2 m VLT telescopes entered service at Cerro Paranal in early 2000, La Silla Observatory's mission changed. The smallest telescopes at the site and the venerable but obsolete photographic Schmidt telescope, long supplanted by electronic cameras, have all been shut down. The 2.2 m and the NTT, which

really led to the development of the VLT, now play supporting roles in identifying the most interesting objects for study by the giant telescopes. The 2.2 m has been equipped with a widefield CCD camera for this purpose, whose 67 million pixels cover a 35 x 35 arcmin area of sky, larger than that of the Moon.

The NTT is now used primarily for deep-sky survey work of dim brown dwarfs, distant galaxies and quasars, which are then studied in more detail with the VLT. One example, the Susi Deep Field Survey, involved a total of 32 hours of exposure with the NTT at four different wavelengths. This survey recorded nearly 500 galaxies fainter than 27th magnitude in a tiny 2.3 x 2.3 arcmin area around the quasar QSO BR 1202-0725.

Many of the other smaller instruments at La Silla also do extended survey and monitoring work requiring tens or even hundreds of nights observing time. This began in the late 1990s with the DENIS and EROS programs and a 1 m telescope. DENIS completed an infrared survey of the southern sky and EROS was an international study of MACHOS, which are low-mass objects in the galactic halo. Not far from the NTT, Swiss astronomers erected a new high-performance instrument, the 1.2 m Leonard Euler telescope, totally dedicated to exoplanet searches. This telescope has discovered more than 20 exoplanets already, including several with Jupiter- and Saturn-like

PROJECT GOODS

The Chandra Deep Field South (CDFS) began in 2000. A survey of a small section of sky in Fornax, involving various space telescopes and the most powerful telescopes in the southern hemisphere, was part of NASA's Great Observatories Origins Deep Survey (GOODS). This survey probed two areas, one in the northern and another in the southern sky. At wavelengths spanning X-rays to radio frequencies, it was at the highest levels of resolution and sensitivity possible. Hubble, Chandra, XMM and SIRFT observed at wavelengths not accessible from the ground, X-rays, ultraviolet and far-infrared. The ESO telescopes involved were the 2.2 and 3.5 m telescopes at La Silla, and the 8.2 m giants at Cerro Paranal Observatory. Visible and near-infrared images of stars and galaxies were obtained approaching 27th magnitude. The VLT telescopes will continue studies of the most interesting objects in the survey, including quasars, active galaxies and galaxy clusters, gravitational lensed objects and very young, highly redshifted galaxies. The most interesting will be examined in detail both by the VLT and the joint U.S.-European millimeter interferometer, ALMA.

■ *The Tarantula Nebula is located in the Large Magellanic Cloud, some 170,000 light-years away. This widefield image, taken with the 2.2 m telescope at La Silla Observatory, is a composite of 50-minute exposures through 456, 540 and 652 nm band pass filters.*

masses. This telescope also discovered the first multiplanetary systems in the southern hemisphere, including two Jupiter-size planets around the star HD82943 in Hydra, and two Saturn-size planets around HD83443 in Vela.

The 3.6 m telescope will also be dedicated to exoplanet research efforts in the coming years, since these require extended and multiple observing sessions. It will be equipped with HARPS for this work, a 100,000 resolution spectrograph suitable for stars as faint as 10th magnitude. This combination will be able to resolve radial velocities of only 1 m/s and detect exoplanets the size of Uranus, Neptune, Saturn and Jupiter to a distance of almost 500 light-years from Earth. This should also make it possible to identify complex planetary systems like our own for the first time. ■

> Coronagraphy

Detecting planets and other dim features adjacent to stars is a very difficult undertaking. Coronagraphs were developed to overcome this problem in solar astronomy by artificially "eclipsing" the Sun's disk and greatly attenuating its intense light. Developed at the beginning of the 20th century, this technique has made tremendous advances since then and today is used to study protoplanetary disks around nearby stars, to search for extrasolar planets and to study regions near the cores of distant galaxies.

SOLAR AND STELLAR CORONAGRAPHY

Because of its low intensity in both visible and infrared light, the Sun's corona could at first be seen only during total eclipses. In the 1920s Bernard Lyot at Pic du Midi Observatory developed the coronagraph, an instrument that blocked out the intense light of the photosphere and

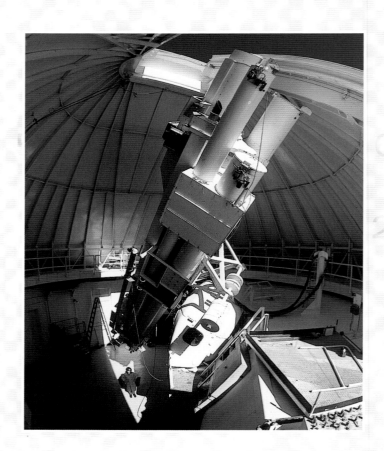

● This big, .4 m coronagraph is located at an elevation of 2,800 m at Sacramento Peak Observatory in New Mexico.

revealed the corona and other solar details. This first coronagraph, called "de Lyot," is still in use today. It consists of a circular focal-plane mask or mirror and a narrow bandpass filter. Most of the sunlight is reflected back, but light from the corona is transmitted around the edges of the occulting mask.

Both ground- and space-based coronagraphs are now widely used to study the corona and the regions adjacent to the Sun. The SOHO spacecraft LASCO coronagraph has been particularly helpful in detecting sun grazers and comets that plunge into the Sun.

More recently coronagraphs have been developed to investigate nearby stars. Although similar to solar coronagraphs, they are considerably more difficult to actually use, since they require much smaller masks that cover only a few arc seconds of sky.

In 1985 American astronomers detected an extended dust ring and possibly the first extrasolar planetary system to be discovered, around the star Beta Pictoris. They used this technique and the 2.5 m telescopes at Las Campanas Observatory. The disk, about a hundred thousand times dimmer than the star itself, extended some 5 arcsec or 100 AU from along its equatorial plane. The question then was, would this technique see detail even closer to this star, as close say as the outer planets in our Solar System? This was accomplished about a decade later with a star coronagraph on the Hubble Space Telescope and with ground-based telescopes using adaptive optics. The Beta

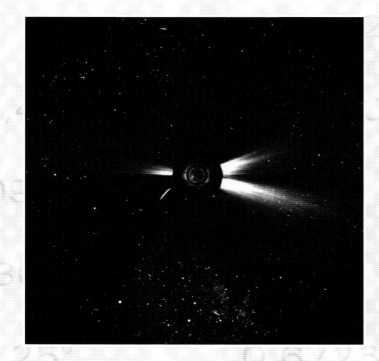

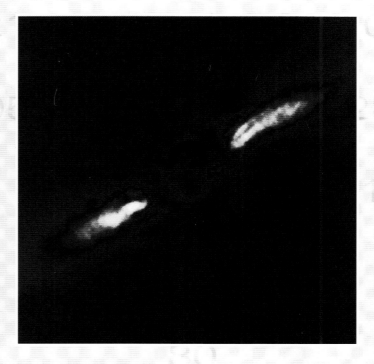

● This LASCO image taken by the SOHO satellite shows the Sun's corona and its surroundings, including a nearby comet. The Milky Way is visible in the background. From this perspective, its stars appear billions of times fainter than the surface of the Sun.

● An image of Beta Pictoris taken at La Silla Observatory with the 3.6 m telescope and the Adonis adaptive optics system. Beta Pictoris is masked in this image to reveal an extensive dust ring some hundred thousand times dimmer than the star itself.

Pictoris dust ring was traced almost four times closer than previously possible.

CONVENTIONAL CORONAGRAPHY

The basic function of a coronagraph is to block just the right amount of light coming from the center of a bright celestial object. This is usually done with a circular mask or an occulting bar at the focal plane of the telescope, although perforated mirrors can also be used to deflect the unwanted light.

Extremely small occulting bars are needed to reveal dust disks and other detail around nearby stars. The star's image must also be sharp and completely stable, because any light leakage around the mask easily overshadows the extremely faint circumstellar detail. Since atmospheric turbulence and bad seeing can impact this type of observation from the ground, it is best done with space-based telescopes.

The most important performance characteristic of a coronagraph is image contrast, which is defined as the brightness ratio between the object being occulted and the detail to be imaged. Since coronagraphy is usually looking at extremely faint detail near very bright stars, the higher the contrast the better. In practice, factors like the type of telescope involved, seeing conditions and image quality, the stability of the adaptive optics, and the type of detector used also play a role.

In most respects modern star coronagraphs perform very well and are sensitive enough to detect giant, still-warm plan-

etary bodies around very young stars. However the ultimate prize still eludes this technology: namely direct visualization of planets around sun-like stars. This will clearly require new types of coronagraphs and technology, something that is being actively pursued at present.

FUTURE DEVELOPMENTS

One new type of occulting mask under development is designed to "null out" starlight by destructive interference. A method called "phase masking" uses a special medium at the focal plane of the telescope to impart a phase shift (B) to half the light in the stellar diffraction patterns, thereby attenuating most of the light by destructive interference. Another method, called "achromatic interference coronagraphy," introduces a similar phase shift (B) using a Michelson interferometer.

These two approaches should theoretically produce more efficient coronagraphs with much higher contrast than existing instruments, particularly in regions closest to stars. Under ideal conditions, this type of coronagraph should detect Jupiter-size planets orbiting sun-like stars.

A process called "nulling interferometry" is also very promising in terms of conventional interferometry. In a few years this approach will be tried from space to search for Earth-like planets around nearby stars. Meanwhile, nulling interferometry will be thoroughly tested with more modest ground-based telescopes. Both the VLT and the Keck interferometers will fully evaluate this approach. ■

The Cerro Paranal Observatory

Except for the intense deep blue sky above, you could easily imagine yourself on Mars. The daytime sky is so breathtakingly clear that you can actually spot some of the brighter planets with the naked eye. The Atacama Desert, a haven of modern astronomy, is an environment of superlatives: the highest, driest, most inhospitable and strangest place in the world. Cerro Paranal is the ideal vantage point to view this magnificent panorama. The peaks and valleys of this mountainous terrain are visible in all directions: yellow- and ocher-colored saddlebacks as far as the eye can see. A mere 12 km to the west, a blanket of clouds perpetually shrouds the Pacific Ocean. The outlines of yellow- and copper-colored hills are visible directly north and south of Cerro Paranal, but lower down and directly west the horizon is marked by a series of increasingly higher mountains of the Andean cordillera. Way off in the distance, some 200 km away, you can spot the snow-covered cone of Llullaillaco; this imposing volcano, reaching 6,778 m above sea level, is the highest point of the Atacama.

CHILE

■ Cerro Paranal rising to an elevation of 2,635 m above the Atacama Desert enjoys more than 330 clear nights a year. The summit is located some 12 km from the Pacific Ocean, which is perpetually shrouded by a cloudy inversion layer.

Here on 2,635 m high Cerro Paranal the elevation and the extreme dryness make it hard to breathe. ESO astronomers discovered this finest astronomical site in the world in 1980, while searching for a location for their planned new Very Large Telescope (VLT). In their worldwide quest for the ideal location, scientists had to develop an entirely new branch of science: astrometeorology. This was required not only for more systematic assessments of astronomical weather conditions, but also to identify locations suitable for work across the full electromagnetic spectrum, including optical, infrared, millimeter and other wavelengths. After an exten-

sive evaluation of sites across the globe, including Réunion Island, the Namibian Desert, southern Africa and the Pamir range, ESO astronomers concluded that the best location anywhere was in the Chilean Andes, especially in the Atacama region. Only Hawaii's Mauna Kea volcano in the northern hemisphere can compare, and it was already home to the world's largest assembly of telescopes.

A .3 m "Seeing Monitor" telescope and a weather station were set up on Cerro Paranal in 1983, to continuously monitor not only weather conditions, but also variations in seeing, nighttime temperatures, wind velocity and humidity for a period of eight years. Perhaps most importantly, sky transparency was also recorded on an hourly basis every night.

Finally in 1991 the ESO directorate decided to move forward with construction of the VLT project at what was considered a near-ideal location. To ensure that this site would remain free of any future encroachment or light pollution, the Chilean government, already a full partner in this European venture, allocated a 725-square-kilometer area of land in and around Cerro Paranal to the observatory.

The two-decades-long record of astronomically favorable weather at the site is astonishing. On average, the site enjoys almost 290 perfect nights a year, nights of "photometric" quality. Another 40 nights are classified as "spectroscopic," meaning that occasional banks of high cirrus clouds hinder observations for a few hours. In short, the Atacama Desert provides more than 330 useful nights annually. Apart from clouds, astronomers are most concerned about the quality of "seeing" or atmospheric turbulence. Here, too, the site excels. Average seeing condi-

■ *The ESO selected Cerro Paranal for the Very Large Telescope due to its remote location and proximity to the Pacific Ocean. The favorable location and exceptional conditions of this site assure that astronomical imaging will be of unsurpassed quality.*

tions are around 0.8 arcsec, with about 150 nights of 0.6 and 0.4 for the top 20 nights a year. This compares with an average of 0.5 arcsec seeing on Mauna Kea, which has considerably fewer nights of photometric quality (230).

The realization of a dream

European astronomers dreamed for nearly a century of being in a position to compete on equal footing with their American colleagues in astronomical research, especially in such prestigious areas as cosmology. Thanks to the generosity of private benefactors in the United States over many years, American astronomers have enjoyed access to some of the most powerful telescopes in the world. La Silla Observatory provided the European astronomical community access to one of the finest facilities beyond the Atlantic. With the VLT project, however, the ESO goal is clearly to challenge America's preeminence with the finest telescopes at the best location on the planet.

First planned in the 1980s and officially begun in 1987, the Very Large Telescope is the most amazing astronomical facility ever conceived. Fully modular in design, the system consists of four separate 8.2 m telescopes, each equipped with three focal-plane instruments permitting sky coverage at wavelengths ranging from 300 nm in the ultraviolet to 25 μm in the infrared, and at several different levels of resolution. No other observatory today has this

■ *The four VLT telescopes can be used both individually and in unison. When joined by three smaller 1.8 m telescopes, the resultant Very Large Telescope Interferometer (VLTI) will have an effective aperture of 200 m.*

range of capabilities. Each of the four telescopes can be used separately or they can be used together to study the same object (supernovae and quasars, for example) with different instrumentation. Eventually the four telescopes will be linked into an array called the Very Large Telescope Interferometer (VLTI), the largest optical and infrared facility of its kind in the world.

To build an array like this, the top of Cerro Paranal had to be lowered by about 30 m and then leveled into a flat surface area of about two hectares. The four VLT domes were sited on the western edge of this platform, facing the prevailing winds, and then linked by underground tunnels housing the interferometer's array of mirrors and cables. The telescopes were given Mapuche Indian names: Antu, Kueyen, Melipal and Yepun — the Sun, the Moon, the Southern Cross and Venus. Each telescope is 20 m long and 16 m wide, and has a movable mass of 430 metric tons. The four Zerodur mirrors were cast by the German company Schott, and polished by the French company Reosc, who nicknamed them Joe, Jack, William and Averell, with reference to the Belgian comic-book characters, the Dalton brothers. Each mirror is 8.2 m in diameter, only .17 m thick and weighs 23 metric tons. With a surface accuracy of 8 nm, the Yepun mirror is considered the finest in the world today. In fact, on those occasional nights when the skies above Cerro Paranal are almost space-like in quality, these telescopes have attained a record 0.2 and 0.4 arcsec resolution even without adaptive optics.

The first VLT telescope, Antu, was commissioned in 1998, followed by Kueyen in 1999, Melipal in 2000, and finally, Yepun in 2001. Antu features three foci: two Nasmyth and a Cassegrain focus equipped with ISAAC (Infrared Spectrometer and Array Camera), a powerful imaging spectrometer operating between 1 and 1.5 μm with infrared spectral resolution of 500 to 3,000. FORS 1 (Focal Reducer and Low Dispersion Spectrograph), installed at the first Nasmyth focus, operates in the ultraviolet to near-infrared range, between 300 and 1,000 nm. Covering a field of almost 7 x 7 arcmin, this instrument is especially well suited for high-quality imaging of galaxies, distant galaxy clusters and quasars. Capable of recording objects as faint as 24th magnitude with exposures of only two minutes, FORS 1 can reach beyond 29th magnitude with very long exposures.

In 2004, CRIRES (Cryogenic High Resolution Infrared Spectrograph) was installed at Antu's second Nasmyth focus. This powerful, next-generation instrument can obtain high-resolution spectra of objects as faint as 18th magnitude, with resolution between 40,000 and 100,000 at wavelengths of 1 and 1.5 μm.

FORS 2, a spectro-imager similar to FORS 1, is installed at the Cassegrain focus of the Kueyen telescope. The first Nasmyth focus supports FLAMES (Fiber Large Area Multi-Element Spectrograph), a versatile, multipurpose device equipped with optical fibers to serve several other instruments. One of these is GIRAFFE, a high-resolution, visible-range spectrograph working between 370 and 900 nm with spectral resolution between 7,500 and 25,000. It can pre-position 130 optical fibers over a 25 arcmin field and measure objects to 22nd magnitude. GIRAFFE will be particularly useful for research on galaxy clusters and gravitational lensing objects. A second set of optical fibers extends some 20 m from FLAMES to Kueyen's second Nasmyth focus, and feed UVES (UV Visual Echelle Spectrograph). The latter provides spectral resolution from 40,000 to 100,000 in the visible range between 300

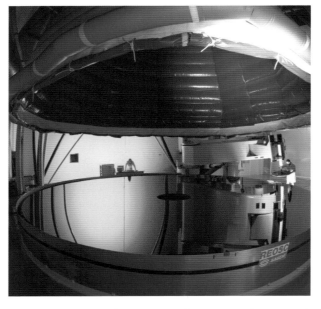

■ *The 8.2 m mirror of the Yepun telescope weighs 23 metric tons. Like all VLT mirrors, it is composed of the ceramic glass Zerodur, and its 50-square-meter surface has been ground and polished to an accuracy of 8 nm.*

and 1,100 nm. Reaching objects to 19th magnitude, UVES is intended for research on stars and the interstellar medium in the Milky Way and other galaxies.

Melipal's two Nasmyth foci are fully equipped with specialized instrumentation; however, its Cassegrain focus can be outfitted with new equipment brought by visiting scientists. This telescope has attained between 0.2 to 0.7 arcsec resolution with its VLT Mid Infrared Spectrometer and Imager (VISIR) operating in the infrared from 8 to 25 μm. In the spectrometer mode, VISIR resolution is between 250 and 30,000. With unrivaled performance in this spectral range, Melipal has been used mainly for research on star formation and protoplanetary systems.

The telescope's second Nasmyth focus supports VIMOS, a special widefield (14 x 14 arcmin) spectro-imager capable of recording up to 750 galaxy spectra in a single exposure. Working in the visible range at resolution levels from 300 to 2,000, VIMOS can record spectra of dim objects to 26th magnitude. As an imager, it can reach objects as faint as 29th magnitude. VIMOS was designed primarily for deep-sky surveys and large-scale sampling of

■ *The spectrographic imager FORS 1 was installed at the Cassegrain focus of the Melipal telescope in 1999. Sensitive from the ultraviolet to the infrared, 300 to 1,100 nm, this sensitive instrument covers a 6.8 x 6.8 arcmin field of view.*

galaxies at cosmological distances.

To complement the optical range capabilities of VIMOS, the Yepun telescope was equipped with NIRMOS, an infrared receiver sensitive in the 1 to 1.8 μm range. With spectral resolution of 2,000, this device can record spectra of 180 galaxies simultaneously, as faint as 23rd magnitude. VIMOS and NIRMOS will be used together to establish distances and physical characteristics of over 150,000 galaxies and to generate a three-dimensional map of the Universe showing objects with redshifts between 5 and 6.

The Cassegrain focus of Yepun supports a receiver called SINFONI (Single Far Object Near IR Investigation). Combined with its adaptive optics system, MACAO, this advanced spectrograph is slated to support other VLT telescopes as well. Working at wavelengths in the 1 to 2.5 μm range and spectral resolution between 1,000 and 4,000, SINFONI is intended for small, angular targets like planetary surface details, galactic nuclei and quasars.

ADVANCED ADAPTIVE OPTICS

The Yepun telescope's other Nasmyth focus supports two advanced instruments. NAOS (Nasmyth Adaptive Optics System) was the first adaptive optics system used by the VLT. Combined with CONICA (High Resolution Infrared Camera), this system has provided near-perfect images in the 1 to 5 μm range. By effectively eliminating the effects of seeing, this combination has attained 0.06 arcsec resolution at 2 μm: something previously attainable only with the Hubble Space Telescope, but at four times shorter wavelengths. This

■ *The Yepun telescope is 20 m long and 16 m wide, with a total movable mass of 430 metric tons. Both the mount and all optics are fully computer-controlled.*

combination of instruments produced some truly astonishing results when first tried in 2001. In a series of test exposures that included the Moon, Jupiter's satellites, the rings of Saturn, as well as stars, nebulae and galaxies, all taken at a wavelength of 1.2 μm, NAOS and CONICA actually exceeded the Hubble Space Telescope by attaining a record resolution of 0.03 arcsec!

In the coming years NAOS and CONICA are expected to break new ground in the areas of planetary system formation, closely-paired red and brown dwarfs, stars at the core of our galaxy, gas envelopes of supergiant stars, and so forth. Due to their extraordinary resolving power and sensitivity, these two instruments will be able to usefully investigate objects as faint as 27th magnitude, including possible direct detection of extrasolar planets.

With its many high-performance support instruments, the VLT system as a whole is without equal at the present time. Never before have astronomers had so much telescope power at their disposal. For starters, the VLT telescopes have a combined collection area greater than 200 square meters. This compares to the 4.5 square meters of the HST. Suitable for work at all levels, from solar-system studies to the most distant quasars, the VLT is poised to address some of the most pressing and exciting questions in astronomy today. The ESO has placed particular emphasis on the detection and analysis of exoplanets, the birth and evolution of stars, and the evolution of galaxies over the 14-billion-year lifespan of the cosmos. This includes deep space surveys of objects with redshifts between 6 and 8: back to the earliest stages of the Universe. The main thrust of this work is to establish firm values for rates of cosmic expansion, as well as the Hubble and cosmological constants.

In addition to providing European astronomers a quartet of 8 m telescopes, the VLT was planned to operate as a giant interferometer. The Very Large Telescope Interferometer (VLTI) will function as a virtual 200 m aperture instrument. When completed, between 2006 and 2010, it will be the most sensitive instrument of its kind in the world. Only the Keck Telescope Interferometer (KTI) in the northern hemisphere will rival it, by combining its two 10 m telescopes and four 1.8 m auxiliary instruments. KTI and VLTI performance levels are expected to be about equal.

THE VISTA TELESCOPE

The Visible and Infrared Survey Telescope (VISTA) will be a very widefield instrument. British astronomers have been planning a telescope of this type for several decades. In the 1970s the 1.2 m Schmidt at Siding Spring Observatory, along with its twins at Mount Palomar and La Silla, conducted comprehensive photographic sky surveys in both hemispheres. With their 5.5 x 5.5 degree field coverage, these instruments made many new discoveries and a catalog of the sky that is still in use today. A consortium of 18 U.K. universities will take that approach to a new level in 2005, when a 4 m digital "super Schmidt telescope" will become operational less than a kilometer from the Cerro Paranal facilities. VISTA will be equipped with the largest electronic camera ever built, with 410 million pixels and an astonishing field coverage of 1.6 x 1.4 degrees in the visible range and 1 x 1 degrees for work between 0.8 and 2.5 μm. This project will map most of the southern sky by covering nearly 10,000 square degrees to magnitude 26. Narrower deep-sky surveys to magnitude 26 and 27 are also likely. The VISTA project brought the United Kingdom into full association with the ESO, which now includes 11 member nations.

The VLTI currently involves the Antu and Melipal 8.2 m telescopes and three movable 1.8 m auxiliary instruments, which can be positioned at 30 different locations on the surface platform. All telescopes are connected by a series of underground mirrors and support instrumentation that links all Cerro Paranal facilities. Initial performance tests with the partial telescope array began in 2001 and the first useful data was obtained in 2002. When completed, the VLTI will operate at 0.6 μm in visible light and at 20 μm in the infrared, using its three main receivers (Amber, Midi and Prima) to synthesize images of various astronomical objects. The VLTI is expected to attain an astonishing 0.001 arcsec at 1 μm and 0.01 arcsec at 10 μm resolution along its 200 m baseline.

The VLTI and KTI's principal strengths are their enormous light-collecting areas. Several other current optical interferometer projects, including CHARA, IOTA, GI2T, COAST, PTI and others based on smaller-aperture telescopes, will be considerably less sensitive. With its combined 8.2 and 1.8 m telescopes and with its anticipated maximum resolving power of 0.0005 arcsec at 0.6 μm, the VLTI will be able to image objects as faint as 20th magnitude and reveal surface detail on nearby stars. Supergiants like Betelgeuse and Antares should reveal as much surface detail as a .25 m amateur telescope does on Jupiter and Saturn. The VLTI will also be able to directly image the most massive and

■ *Spanning some 800,000 light-years, NGC 6872 is one of the largest galaxies known. This immense spiral and its smaller companion IC 4970 are located 300 million light-years from us. This image of the pair, taken with the FORS 1 spectral-imager at the Melipal telescope, is a tricolor composite taken through blue, yellow and red filters.*

warmest exoplanets, and indirectly detect terrestrial planets around stars between 10 and 20 light-years from us. With resolution of about a billion kilometers near the center of our galaxy, the VLTI will be able to observe regions adjacent to the massive black hole at its core. It will also be possible for the first time to scrutinize the nuclei of other galaxies in detail, as well as the centers of quasars. For instance, the VLTI should reveal detail at the heart of the well known quasar 3C273, located about two billion light-years from Earth, as if it were a mere 10 light-years away.

The VLT project has cost the multinational ESO consortium less than US $372 million and is expected to provide high-quality results beyond 2020, after which it is likely to replaced by even more powerful telescopes. The current ESO program is expected to lead to further advances in adaptive optics and interferometry for the next generation of giant telescopes. A much larger-aperture instrument is already in the planning stages. This OWL (Overwhelmingly Large) telescope will be 100 m in diameter. ■

The Siding Spring Observatory

With only 3,000 inhabitants, the sleepy hamlet of Coonabarabran basks in the perennial blue skies of New South Wales. With its long, wide streets, its churches and its many pubs, the town is the last stop for truckers enjoying their Toohey's on tap before continuing their long trek to Queensland. Drawn by good weather and clear skies, many amateur astronomers come here for their annual vacations. Coonabarabran is one of the few oases in this isolated region of the great island continent of Australia. Fourteen times the size of France, Australia has less than 20 million inhabitants, most of whom live on its southwest coast in cities like Sydney, Melbourne, Adelaide, Brisbane and Canberra, the small national capital.

Most of the technical and engineering staff at Siding Spring Observatory lives in Coonabarabran, only 30 km from their place of work. Driving along the road to Coonamble, they can count themselves fortunate to be

■ The 3.9 m telescope at Siding Spring is one of the most productive contemporary instruments. Australian and U.K. astronomers still actively use this facility, established in 1975.

working at this Anglo-Australian facility. Sited atop a eucalyptus-covered ridge, Siding Spring Observatory overlooks Warrumbungle National Park, one of the most beautiful natural preserves in Australia and a sanctuary for some of the most unusual fauna and flora in the world. From Coonabarabran itself, the 50 m dome of the observatory's largest telescope looks like a tiny white pearl, shining brightly against the blue sky.

Both majestic Warrumbungle and Siding Spring Observatory are vital economic assets for this quiet region, which draws some 100,000 tourists each year from the coastal cities up to 500 km away. Australians are very proud of their national achievements

in science and research, and this is clearly reflected in their huge investment in astronomy and space science. For example, although they represent a scant 0.3 percent of the world's population, Australian scientists publish almost 3 percent of all professional research papers in astronomy and astrophysics!

At an elevation of only 1,150 m above sea level, Siding Spring is unlike most of the larger modern observatories in the Andes or on Mauna Kea. Instead of cold desert locations with rarefied atmospheres, tall eucalyptus trees swaying in the wind surround the observatory. These trees cast long shadows onto the domes of the observatory at dusk, amid a cacophony of colorful wild parrot shrieks. Still, in a country whose highest peak, Mount Kosciuszko, is but 2,228 m above sea level, Siding Spring is clearly a mountain. Occasionally in the middle of winter in mid-August, the trees get a dusting of snow. Far from city lights and air pollution, and protected from dust by the surrounding greenery, this Anglo-Australian observatory enjoys exceptionally dark, clear skies.

Although many of the most interesting and important celestial objects are found in the southern hemisphere, they were historically all but ignored by northern observers. The fact that this situation is largely reversed now is due in large measure to the work of British and Australian astronomers at Siding Spring, still one of the most important centers of astronomical research in this hemisphere.

Siding Spring Observatory developed in two distinct phases. The first, totally Australian undertaking began with the Australian National University (ANU). In the 1950s, ANU oversaw Mount

Stromlo Observatory near Canberra. This important institution was established in 1924 and in 1955 acquired the famed 1.9 m telescope, which

■ *The equatorially mounted, 3.9 m AAT is equipped with some truly advanced instrumentation. This includes the UHRF spectrograph, featuring a record spectral resolution of 1,000,000 for really bright stars, and the 2dF spectrograph, which can record spectra of 400 faint galaxies during a single exposure.*

finally opened a window to the southern skies. This telescope and its twin at Radcliffe Observatory in Pretoria, South Africa, were the most powerful southern instruments at that time. This was a frustrating situation, however, since all giant telescopes, including Mount Palomar, Wilson and Hamilton, were sited in California and not in a position to study the exceptional riches of the southern skies. Objects like the Magellanic Clouds (ten times closer to us than the Andromeda Galaxy), Omega Centauri and 47 Tucanae (the largest and closest globular clusters), and the very center of the Milky Way were all simply not getting the attention they merited.

THE ANGLO-AUSTRALIAN TELESCOPE

Eventually such well-known American researchers as Bart Bok and Gérard de Vaucouleurs joined Mount Stromlo Observatory to investigate the Milky Way and the Magellanic Clouds in more detail. However, when lights from the growing national capital,

Canberra (which now has more than 250,000 inhabitants), started to seriously affect observing conditions at Mount Stromlo, the ANU began the search for a really dark, pollution-free site in a less populated area of Australia. And so Siding Spring was born. A 1 m telescope was installed there in 1964 and is still in active use today. About 20 years later, the largest national telescope was built there, the 2.3 m Advanced Technology Telescope (ATT). Altazimuth-mounted and fully computer-controlled, this revolutionary telescope featured two Nasmyth foci as well as a traditional Cassegrain focus. The design of this telescope and its simple, movable dome served as a model for similar instruments elsewhere, including the European New Technology Telescope.

The second phase of development at Siding Spring was international in scope. In the 1950s only the United States was home to truly large telescopes, a lead it maintained to the end of the millennium. No other major country came anywhere close. Just as the Australian government began construction of Siding Spring Observatory, several other nations, particularly in Europe,

started planning for a new generation of telescopes that would be smaller and less costly than Mount Palomar but just as powerful, thanks to major advances in technology.

After many years of planning and negotiating, the Australian and British governments signed an agreement in 1967 for a jointly run observatory at Siding Spring. An equatorially mounted 4 m class telescope was selected, similar to the ones planned for Kitt Peak and Cerro Tololo, and in 1975 the 3.9 m Anglo-Australian Telescope (AAT) was placed into service. At the time the AAT was the most powerful telescope in the southern hemisphere and the third largest in the world.

From its very beginning the AAT produced some outstanding astronomical images and thanks to a battery of first-class support instrumentation, quickly earned universal recognition for the quality of research carried out there. With foci of 12, 31, 58 and 149 m, the AAT can support a broad range of instruments. The primary focus (accessed via an observer's cage at the top of the telescope tube and just a few meters below the observatory dome) was used extensively for photography by David Malin. This well-known English astronomer developed novel photographic techniques in the 1970s, not only to enhance images but also to generate true color renditions of astronomical objects. He also applied these techniques to photographs taken with Siding Springs' Schmidt telescope.

The Anglo-Australian Schmidt is a twin of the famed Palomar Schmidt. Both are in effect huge telephoto lenses of 3,000 mm focal length and f/2.5 focal ratios. In addition to undertaking a systematic survey of the southern sky, David Malin discovered a new class of faint dwarf galaxies with this instrument, as well as the tenuous gas that envelopes many giant galaxies.

The Schmidt also played a major support role for the AAT. Because of its wide 4 x 4 degree field coverage, this instrument recorded countless distant galaxies and quasars, which were then studied in detail spectroscopically with the AAT.

The AAT has undergone several upgrades since 1975. Obviously its photographic plates were long ago replaced by much more sensitive CCD cameras. Thanks to Australian expertise in fiber

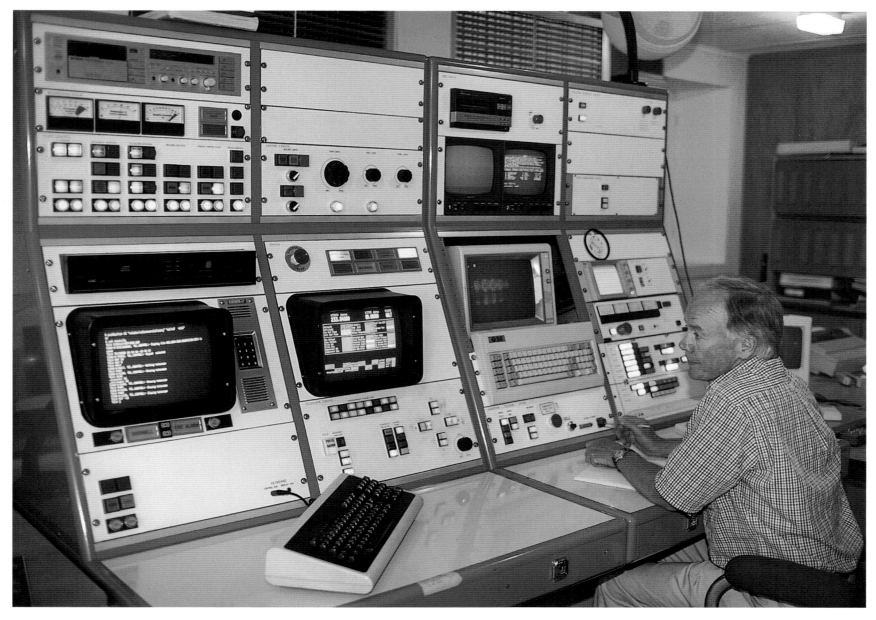

■ *The control center of the Advanced Technology Telescope (ATT) is shown above. When placed into service in 1984, the ATT was truly a cutting-edge instrument. Weighing less than 30 metric tons, this compact, altazimuth-mounted telescope boasted an ultrafast f/2, 2.2 m mirror.*

optics and electronics, entirely new support instrumentation has also been developed for this telescope, including the Ultra High Resolution Facility (UHRF), the most powerful spectrograph ever built. Mounted at the f/36, 140 m Coudé focus, the UHRF can attain a record-high spectral resolution of 1,000,000 with stars brighter than 4th magnitude.

SEEING THE UNIVERSE IN 3D

A second extremely sensitive spectrograph known as UCLES (University College London Coudé Echelle Spectrograph) was installed in 1998 and used by teams of Australian, British and American astronomers for systematic searches of extrasolar planets. This bore fruit very quickly. Only three years later, Jupiter-size planets were discovered around Mu Ara, Epsilon Reticulum and HD179949 in Sagittarius.

Without doubt, however, Siding Spring astronomers have garnered most attention through their 2dF (Two Degree Field) survey. Initiated in 1995, this program used a highly innovative spectrograph at the prime focus of the 3.9 m telescope. Working with fiber optics, the 2dF spectrograph recorded 400 spectra of faint objects simultaneously in a 2 x 2 degree field. By 2002 this program had completed the largest deep-sky survey ever undertaken; it recorded spectra of over 230,000 galaxies and more than 20,000 quasars to 20th magnitude. Covering an area of the sky of almost 1,700 square degrees, the survey provided a three-dimensional map of the Universe out to almost 10 billion light-years and showed that the large-scale distribution of galaxy clusters outlines a network of filaments and voids whose average diameter is about 300 million light-years. The data also indicate that the average concentration of matter in the cosmos is just over 1 atom per cubic meter. This translates into a density ratio of $\omega = 0.3$, in agreement with cosmological models that suggest that the Universe is open and its expansion is accelerating due to enigmatic "dark energy." Predicted by Einstein's "cosmological constant" nearly a century ago, this expansion implies that galaxies will continue to recede for billions of years to come. ■

The Parkes Radio Observatory

As you make your way north along the Newell Highway on the way to Dubbo, amid fields of wheat as far as the eye can see, with sheep grazing in the sun-drenched prairie, you are struck most of all by the utter silence. Sydney is 365 km behind you and the town of Parkes lies 25 km ahead. As with most of Australia, this region of New South Wales is very sparsely populated. When you reach the Gooban Valley you are greeted by an imposing radio telescope, clear evidence of Australia's vibrant contribution to astronomy. The ribbed metal dish of this truly eye-catching instrument seems almost transparent. Fully steerable in every direction, the telescope is perched atop a three-story tower containing control rooms, computers, data-storage facilities and accommodations for engineers, technicians and astronomers. The telescope operates around the clock every day of the year. Weighing more than 1,000 metric tons, the 64 m diameter Parkes dish has a total collecting surface of 3,000 square meters. Except for the 70 m NASA deep-space tracking antenna at Tindbinbilla, Parkes is the largest radio telescope in the southern hemisphere.

■ Inaugurated in 1961, the imposing Parkes radio telescope has enjoyed an exceptionally long scientific life. This 64 m diameter Australian radio dish is still the largest in the southern hemisphere.

Shortly after entering service in 1961, the Parkes radio telescope made one of the most important astronomical discoveries of the 20th century by helping clarify the nature of the mysterious objects called quasars. The first of these unusual objects, 3C273, had been discovered just two years before at Cambridge, England, and posed an immediate dilemma. Although this was one of the most powerful radio sources in the sky, optical telescopes detected nothing unusual in Virgo where the signal originated: no unusual stars, nebulae or galaxies could be linked to it. As more radio sources like 3C273 were found, a new term was coined for these mysterious objects: Quasi Stellar Astronomical Radiosources, or quasars. Quasars proved very perplexing for astronomers at the time. Radio telescopes lacked the resolving power to pinpoint their locations accurately enough to be observed with optical telescopes. So where and what were these enigmatic objects?

On August 5, 1962, astronomers at Parkes took advantage of a relatively rare opportunity to observe 3C273 just as it was occulted for about one hour by the Moon. This would at last narrow its position to a defined area of sky. The data that astronomers Cyril Hazard and John Bolton obtained was good enough to pinpoint the quasar to within 1 arcsec, about a thousand times better than was ordinarily possible with the Parkes telescope. A search of Palomar photographic sky survey plates showed 3C273 as a faint, blue-colored object of 13th magnitude with an odd jet-like streak. A few months later, Dutch astronomer Maarten Schmidt obtained a high-quality spectrum of this object with the 5 m telescope at Mount Palomar. With six intense emission lines, something never before encountered, the spectrum was all but uninterpretable. After several months of analysis, Schmidt realized in early 1963 that the emission bands were in fact due to ordinary hydrogen, only they were shifted 16 percent into the red end of the spectrum (redshift of 0.16). Converting this into universal recession rates indicated that far from being a dim star in our own galaxy, 3C273 was not only over 2 billion light-years from us, but emitted as much total energy as 100 galaxies like ours!

While justifiably pleased with their contribution to the

■ *Over the years Parkes Radio Observatory has discovered quasars at ever-farther distances from us: PKS0106+01 with a redshift greater than 2 in 1966; PKS0237-23 with z = 2.22 in 1967; and in 1982 one of the furthest objects known: PKS2000-33t0 with z = 3.78.*

clarification of the nature of the first quasar, Australian astronomers at the time were also aware that they could not have finalized the discovery of this object, since they lacked optical telescopes powerful enough to do so. A few years later, construction began on Siding Spring Observatory.

After the discovery that quasars were distant objects at the edge of the Universe, U.K. and Australian astronomers dominated the field of radio astronomy and ushered in the era of observational cosmology. Since no other comparable instrument has yet been built in the southern hemisphere, the Parkes radio telescope has enjoyed unfettered access to the southern skies and discovered almost half of all quasars known to date.

In addition to its longstanding collaboration with radio astronomers from the United Kingdom, Parkes researchers teamed with American astronomers in an effort to map the entire southern sky at a wavelength of 6 cm. Using several telescopes at both equatorial and southern latitudes, this effort ultimately identified 36,640 distinct radio sources.

Following successful termination of that project, Parkes researchers undertook a special project of major interest to cosmologists. Neutral hydrogen is not detectable by optical means but has a distinct radio signature at .21 m. Although neutral hydrogen is relatively thinly distributed, it most likely accounts for most baryonic (visible) matter in the Universe. To facilitate this survey and complete the task within a reasonable period of time, a special widefield receiver was built in 2000 with multiple-band pass capability. In addition to mapping the distribution of hydrogen gas in the Milky Way, HIPASS will also be used to search for galaxies that are hidden from view by the massive number of stars in our own galaxy. Parkes Observatory is likely to discover thousands of such objects.

Finally, Australian and U.K. astronomers are also mapping neutral hydrogen distribution in galaxies with apparent recession velocities of less than 15,000 km/s, namely galaxies about 150 million light-years from us. Except for the brightest galaxies, this region of the local Universe has not been well described, particularly in terms of dwarf galaxies and other large but extremely faint objects. It's been suggested that tens of billions of such objects may be scattered throughout the Universe, but like the recently discovered galaxy Malin 1, have gone largely undetected. This huge object, discovered at Siding Spring Observatory, is ten times larger than the Milky Way, yet was only recently discovered due to its extremely low surface brightness. If indeed tens of billions of such neutral hydrogen clouds are scattered among galaxies, it implies that much of the missing mass in the Universe has existed in this form from the very beginning. ■

> Astrometry

The branch of astronomy dealing with the measurement and calculation of the position and motion of celestial objects is broadly called astrometry. Technological advances through the ages and now space-based observations have combined to make this an increasingly more precise undertaking. As a result, our understanding of the workings of the Universe as a whole has also increased greatly.

THE IMPORTANCE OF ASTROMETRY

Accurate information on the position, speed and motion of celestial bodies, whether planets, stars or galaxies, is clearly of great relevance to understanding their origin, evolution and ultimate fate in both space and time. Such information is important not only to model various cosmic processes but also to be able to extrapolate from that to the Universe as a whole. Positional astronomy is really a cornerstone of modern cosmology, since it helps define the multidimensional nature of space and time.

ASTROMETRIC METHODS

Three coordinates normally define the position of any object in the sky. These are not absolute positions, but are defined with respect to fixed reference points. Although we know that nothing is in fact "fixed" in space, many objects are so far away that their apparent motion is too slow to be measured with

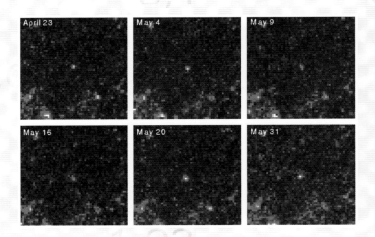

● This series of Hubble Space Telescope images shows variations in brightness of a Cepheid variable in M100, a galaxy about 50 light-years away. Astronomers use variable stars like this, with well-defined periods and luminosities, as reliable distance indicators.

current instruments. Objects like that are used as "absolute" reference points. The position of such objects is typically known to an accuracy of a thousandth of an arc second, as determined by long baseline radio interferometry.

Several positional measurements are generally required to determine the proper motion and velocity of objects in space. This is usually accomplished by taking a series of measurements of its position over an extended period of time. Obviously this becomes more difficult and less reliable the farther away the object is from us.

The parallax method is used to measure distances to stars that are comparatively close to us. This is done by first determining the star's position relative to background stars and then again six months later when the Earth is at the opposite point in its orbit around the Sun. The star's apparent displacement, or parallax angle, is then used to calculate its distance by triangulation. Since parallax angles are always very small, this method only works for relatively close objects. Distance

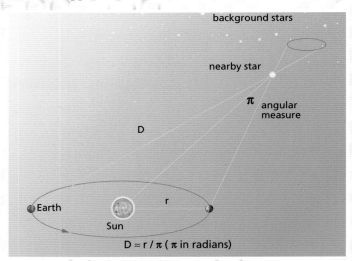

● This diagram illustrates the parallax method of determining the distance of a relatively close star. By taking two measurements of its position relative to background stars, six months apart, when the Earth is at very different points in its orbit, the star's apparent angle of displacement is determined and its distance calculated by triangulation.

● The European astrometric satellite Hipparcos obtained brightness, positional, parallax and proper-motion measurements of more than 118,000 stars to a distance of nearly 3,000 light-years. It successor, Gaia, slated for launch in 2010, will obtain a thousandfold more precise measurements out to 100,000 light-years. This will allow us to determine the exact distance of globular clusters like M3, imaged here with the CFH12K camera and the Canada–France–Hawaii Telescope.

determinations to far away objects are usually based on more indirect methods. Photometry, for example, can be used if the object in question is well understood and its absolute brightness known. Another indirect method is based on Cepheid variables: these are variable stars whose period and luminosity are used to determine their absolute magnitudes, which then serve as reliable distance indicators. Indirect methods of distance determination are obviously less accurate than direct methods, since they depend not only on the quality of the photometric data obtained but also on a clear understanding of the object in question.

ASTROMETRIC INSTRUMENTS

The astrolabe was the first instrument developed for positional measurements of stars and other celestial objects, but it was eventually replaced by transit telescopes. Measurements obtained this way were accurate to about a tenth of an arc second. Astrometric accuracy improved significantly with the introduction of optical and radio interferometry, but space-based telescopes have revolutionized the process completely. Operating from 1990 to 1993, the astrometric satellite Hipparcos obtained parallax (and hence distance), positional and apparent velocity measurements of more than 118,000

stars to the unprecedented accuracy of 0.002 arcsec. This level of precision was anywhere from ten to a hundred times better than previously achieved, and ten thousand times better than that obtained in 100 B.C. by the satellite's namesake, the Greek astronomer Hipparcos! In 2010 the Gaia astrometric satellite is expected to obtain measurements down to a few millionths of an arc second. It will not only map distances and positions of tens of millions of stars in our galaxy with unprecedented accuracy, but also stars in nearby galaxies. ■

● Supernova explosions, which can be brighter than several billion suns for a few days, are used as distance indicators to very remote galaxies. The supernovae (arrows) shown in these HST images are more than five billion light-years distant.

The Narrabri Interferometer

Everywhere you drive through the Australian bush on the long, monotonous route from Coonabarabran to Narrabri, road signs caution you to look out for wildlife. Though rather unnerving at times, this advice is well worth heeding, particularly after you almost hit a big gray kangaroo that suddenly emerged from the side of the road. These familiar, oddly human-looking animals are nearly 2 m tall and seem completely oblivious to the cars and trucks trying to avoid them. A little farther along, an emu as big as an ostrich darts across the road followed by a couple of chicks the size of a Christmas turkey.

Just like the farmers and ranchers of the region, the scientists and technicians at the Narrabri radio observatory drive cars and trucks equipped with "bull or roo bars" — big, ugly front-end bumpers. Designed to fend off kangaroos and other animals so common in this region of New South Wales, these odd-shaped contraptions look something straight out of *Mad Max*.

Located on several dozen hectares of bush, the Narrabri Observatory is a veritable refuge for countless species of marsupials and other animals whose tracks are everywhere. The animals have long grown accustomed to the noise and movement of the giant metal structures that occupy this site. The Narrabri interferometer consists of six identical 22 m radio telescopes connected by a cable network of optical fibers. One of the antennas is fixed in place, but the other five can be moved anywhere along a 3 km stretch of railway tracks. This is the most powerful radio interferometer in the southern hemisphere. Like the VLA, its larger counterpart in the northern hemisphere, the Narrabri interferometer was initially

■ *The Narrabri interferometer sits amid eucalyptus trees in the Australian bush. The six 22 m dishes can be moved on rails anywhere along a 3 km long baseline.*

designed to observe in the wavelength range of 8 to .21 m. Due to its remarkable capabilities, however, this Australian facility has undergone several upgrades and improvements since 1990, making it far more powerful today than initially planned.

By greatly improving the surface accuracy of each dish, installing ultrasensitive new detectors and replacing connector cables with optical fibers, the interferometer's range and sensitivity now far exceed original expectations. In 2001 four of the six telescopes began observations in the 3 to 12 mm range, and by 2004 the full array had that capability. While providing very high-resolution images of nebulae and galaxies, the array's main strength is in the submillimeter range.

While the Narrabri facility is clearly a unique astronomical instrument on its own, it also the core of a still larger interferometer, the Australia Telescope. This immense array extends over the entire Australian continent and can receive radio waves between .3 and .21 m. In addition to the six 22 m Narrabri dishes, the array is linked with a 22 m dish at Mopra near Siding Spring, the 64 m Parkes telescope, the 30 and 70 m NASA dishes at Tindbinbilla, a 30 m telescope at Ceduna, a much smaller 15 m instrument near Perth and a 26 m dish near Hobart, Tasmania. This gigantic array, whose maximum extent is 3,000 km east/west, is the southern hemisphere's counterpart to the American VLBA.

THE SKA MEGATELESCOPE

The principal research focus of the Narrabri array and the extended Australia Telescope has been the evolution of pulsars, the changing

■ Thanks to their experience with the Narrabri interferometer, Australian astronomers are deeply involved in an international project to construct a Square Kilometer Array (SKA), a giant radio telescope with a collecting surface of one million square meters.

aspects of supergiants and the birth of stars and protoplanetary systems in dark nebulae. In 2000 Australian astronomers also began to investigate objects recorded in the Hubble Deep Field Survey. Due to its high sensitivity and spatial resolution at radio wavelengths comparable to the HST in the optical range, the Australia Telescope extended the Hubble survey into other areas of the electromagnetic spectrum. In 2001 this led to the discovery of "source C," an extremely distant galaxy rich in gas and dust. This object, first detected in the 100-hour-long HST Toucan Deep Field Survey, is barely visible at optical wavelengths but stands out prominently at radio wavelengths. Though too faint for spectroscopic analysis by even the biggest existing optical telescopes, source C appears

■ Modeled after Merlin, the VLBI and the VLBA, the Australia Telescope is a giant interferometer. Combining some 13 radio telescopes, the array extends across the entire Australian continent.

to lie anywhere from 6 billion to 12 billion light-years from us. The energy output and distance of this galaxy indicates that it was at an active star-forming stage in the very early Universe, suggesting that star formation began only a few million years after the big bang.

It is clear, however, that far more powerful and sensitive radio telescopes are required to study such extremely faint and distant objects in more detail. That is why Australian astronomers have joined a multinational effort to build the next generation of radio telescopes, including the Square Kilometer Array (SKA), slated for completion in about 12 years. With a one-square-kilometer collecting area, this "megatelescope" should provide direct access to the earliest hydrogen-rich structures in the primordial Universe. ■

The Narayangaon Interferometer

INDIA

They look like huge flowers on some alien landscape, delicate and translucent in the hot tropical twilight. Tall and regal, the 10 radio telescopes are planted amid wheat fields and sugarcane plantations near the small town of Narayangaon. Distributed over an area of several hundred square kilometers, about 250 km east of Bombay and 80 km north of Poona in the state of Maharashtra, the Narayangaon interferometer is one of India's most ambitious scientific undertakings. Officially called the Giant Meterwave Radio Telescope (GMRT) by the National Center for Radio Astrophysics (NCRA) at the University of Poona, the array is also one of the most powerful interferometers in the world. Like the American Very Large Array (VLA), this Indian facility is laid out in a Y-shaped pattern, but is larger overall. The GMRT can deploy as many as thirty 45 m dishes in the Y configuration along a maximum baseline of 25 km. The dishes are fixed in place, mounted on concrete towers and linked underground through optical fibers.

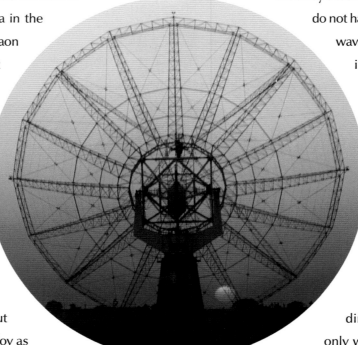

■ *The Giant Meterwave Radio Telescope (GMRT) is one of the largest astronomical instruments in the world. Extending over 25 km near the Indian town of Poona, this radio interferometer consists of 30 dishes, each 45 m in diameter.*

In addition to its sheer size, the GMRT array is also unusual in that it is sensitive to wavelengths ranging from millimeters through to centimeters and decimeters. Unlike most other radio telescopes in the world, however, this Indian facility is designed to observe in the meter wavelength range as well. Since entering service in the late 1990s, the GMRT has been equipped with a series of antennas and receptors sensitive to .21, .5 and .9 m, as well as 1.3, 2 and 6 m wavelengths. These combinations are of major scientific interest; the GMRT is the only instrument capable of probing the sky in this portion of the radio spectrum.

Economic considerations also played a role in designing the GMRT. One advantage of working at longer wavelengths is that the array's technical specifications (and hence cost) do not have to be as stringent as with shorter wavelengths. A 45 m radio dish working in the millimeter range requires a highly reflective surface accurate to a tenth of a millimeter or less. The same dish working at meter wavelengths only needs surface accuracies to a few centimeters. The Narayangaon dishes are parabolas of metal lattice construction mounted on extremely lightweight support arms. Thus, despite their large dimensions, each radio telescope only weighs 120 metric tons in contrast to the 235 metric tons of each 25 m VLA dish and the more than 1,000 metric tons of the 64 m Parkes radio telescope.

The GMRT can attain 2 arcsec resolution at .21 m and 1 arcmin at 6 m; the latter is equal to the resolving power of the human eye. While this level of resolution seems modest compared to what is possible today at much shorter wavelengths, it is most impressive. The Indian array really stands out in terms of its overall collecting surface. Equivalent to a single 250 m dish, it is exceeded only by the 300 m Arecibo radio telescope. Because of this, researchers at the National Center for Radio Astrophysics are naturally concentrating on very weak celestial radio sources, sources that are beyond the

■ *This four-hour exposure by the Hubble Space Telescope of two galaxies in Coma Berenices was taken with the ACS camera. Known as NGC 4676, this pair of gravitationally interacting galaxies is 300 million light-years distant and will merge into one giant galaxy in several hundred million years.*

■ *During each space shuttle mission to the Hubble Space Telescope, astronauts brought the telescope into the craft's service bay for repairs and installation of improved cameras and spectrographic equipment. Hubble orbits the earth at an altitude of about 600 km.*

The Hubble Space Telescope

On a beautiful moonlit night on March 1, 2002, seven astronauts aboard the space shuttle *Columbia* were launched into a 600 km high orbit for a service mission to that most famous of all astronomical instruments, the Hubble Space Telescope (HST). Several similar space missions had preceded this one. *Discovery* had placed Hubble into orbit on April 24, 1990, and *Endeavor* followed on December 2, 1993, for the first repair mission. *Discovery* undertook the two subsequent missions, February 11, 1997, and December 19, 1999. Another visit is scheduled for 2006.

The extraordinary history of the space telescope began in the late 1970s, when NASA and the European Space Agency (ESA) decided to develop an orbital optical telescope that would be free of atmospheric interference and turbulence. The existing 3 to 5 m ground-based telescopes at the time simply could not provide better than 1 arcsec resolution because of atmospheric interference, a level of resolution some thirty times lower than their theoretical limit. This convinced astronomers 25 years ago that the only way to circumvent this was to go into space. Today, thanks to great advances in optical electronics and the establishment of new observatories at high-altitude desert sites, this is no longer as pressing an issue, particularly as modern, ground-based instruments approach performance levels comparable to orbiting observatories.

Equipped with a 2.4 m diameter primary mirror, the HST can observe across a wide range of wavelengths, including ultraviolet, visible and near-infrared radiation from 120 nm to 2.2 μm. Although it barely fit into *Discovery*'s cargo bay during the launch, the 12-metric-ton instrument measures 14 x 12 m when its two large solar panels are fully extended. The HST primary mirror weighs

■ *Placed into orbit in April 1990, the HST has since been serviced four separate times by NASA and ESA astronauts. An advanced new camera, WFPC3, was to have been installed in 2004, but now awaits a future service mission.*

only 740 kg, has a classic f/24 Cassegrain configuration and is supported by five instrument banks that can be used concurrently. Since regular servicing was planned from the start, all instruments are modular units and readily exchanged.

Well before Hubble was placed into its 600 km high orbit above all the haze, dust and turbulence of the Earth's atmosphere, it was expected to totally outperform the largest ground-based telescopes then available. It would be nearly a hundred times more sensitive, attain from 0.05 and 1.0 arcsec resolution and a limiting magnitude of 29 to 30. Although construction began in 1977, the process was beset with so many technical and design difficulties that the initial 1983 launch date had to be delayed. Things were put on hold indefinitely in 1986 when NASA suspended all regular shuttle flights after the *Challenger* disaster, in which seven astronauts perished. Since no other spacecraft was capable of launching such a heavy satellite, another four years passed before the HST was finally carried aloft on April 24, 1990, by the shuttle *Discovery*.

A very nasty surprise, however, awaited researchers who carried out the first HST tests: the telescope had faulty optics! An error in the grinding process of the primary mirror had introduced severe spherical aberration. In the following years, while limited observations were carried out at much-reduced performance levels, the solution to the problem was worked out. This resulted in a system called Corrective Optics Space Telescope Axial Replacement (COSTAR), which was designed and built for installation during the first HST service mission. Although image corrections and other adjustments were applied to compensate for Hubble's optical defects, it became clear that the only way to render this $3 billion instrument fully operational was through additional and costly repairs. Fortunately this was successfully accomplished in 1993, during a service mission by the space shuttle *Endeavor* and its crew of seven.

When initially launched the HST was equipped with a Wide Field Planetary Camera (WFPC), a Faint Object Camera (FOC), a High Speed Photometer (HSP), a High Resolution Spectrograph (HRS) and a Faint Object Spectrograph (FOS). The WFPC was replaced during the repair mission with the more advanced WFPC2 camera, which went on to produce the many extraordinary images Hubble has taken since then. The COSTAR system replaced the never-used original photometer, thereby making it possible to properly operate the European FOC and the two spectrographs.

Several new instruments were installed during subsequent service missions, including the Near Infrared Camera and Multi Object Spectrometer (NICMOS) in place of the original spectrographs, a new Space Telescope Imaging Spectrograph (STIS) in 1997, and in 2002, the Advanced Camera for Surveys (ACS) to replace the FOC. In 2005, the WFPC2 is slated to be replaced by the more advanced WFPC3, and the COSTAR by an ultraviolet Cosmic Origins Spectrograph (COS).

After its optical defects were corrected in 1993, the HST was finally able to undertake its primary mission: exploring the cosmos with unprecedented clarity. In the following dozen years its performance was nothing short of spectacular, despite the fact that larger ground-based telescopes with adaptive optics had been developed. With its new ACS camera, Hubble was able to record objects to 28th magnitude in four hours and to 30th magnitude with 30-hour exposures. Moreover, it has done this at levels of resolution still difficult to attain even with larger Earth-based instruments. Hubble's WFPC2 camera, for example, provides 0.1 arcsec resolution over a slightly constrained 150 x 150 arcsec field of view. By far the highest performing HST instrument is the ACS camera, sensitive from the ultraviolet to the near infrared, between 200 nm and 1.1 μm. In its

■ *In 1994 Hubble was able to provide the first direct image of the surface of another star, Betelgeuse in the constellation Orion. The apparent diameter of this supergiant is 0.056 arcsec. To obtain this historic image, the European FOC camera was used at f/151, which equates to a focal length of 362 m. This hour-long exposure, taken in ultraviolet light at 253 and 280 nm, attained a record-high resolution of 0.03 arcsec.*

wide-angle mode of 205 x 205 arcsec, the ACS with its 2,048 x 4,096 pixel CCD array can attain a resolution of 0.05 arcsec, and 0.025 arcsec with its second array and a 25 x 25 arcsec field of view. When it is installed in 2005, the more advanced WFPC3 camera, boasting a CCD chip of more than 16 million pixels, should provide 0.04 arcsec images in a 160 x 160 arcsec field.

Hubble has established an outstanding record of scientific discovery in all areas of astronomy, extending from the Solar System to the furthest reaches of the cosmos. To put it in perspective, in its first dozen years of operation Hubble logged more than 300,000 observations involving more than 30,000 celestial objects, leading to more than 3,000 scientific publications. Among these are several historic firsts. While operational between 1990 and 2002, the European FOC camera provided the first direct image of the surface of another star, the red supergiant Betelgeuse. Despite its narrow 14 x 14 arcsec field of view, this camera also attained a record 0.03 arcsec resolution in the ultraviolet, and obtained the first detailed images of the surface of Pluto and the first images of brown dwarf stars and the nuclei of galaxies.

Hubble's WFPC2 camera, NICMOS, and the spectrograph-imager STIS have both provided images and data of unprecedented clarity and detail, especially of embryonic star systems and proto-planetary disks. These instruments have also yielded extensive data on stellar populations in open and globular clusters, as well as on planetary nebulae in the Milky Way and other galaxies.

Without doubt, however, Hubble's most important contributions have been in the realm of cosmology, particularly by establishing reliable indicators of the age of the Universe. This is a question researchers have been grappling with for some time.

The apparent recession velocity of galaxies is an index of the rate of expansion of the Universe since the big bang — in other words, as a measure of its age. These parameters also define the Hubble constant,

■ *The Hubble Space Telescope has benefited greatly from the spectacular advances in instrumentation. For instance, WFPC2, the second-generation camera installed on the HST in 1993, was equipped with a 2 megapixel chip while the ACS installed in 2002 contained a 16 megapixel chip.*

named after American astronomer Edwin Hubble, HST's namesake, who formulated it in the 1920s. Use of the term "constant" in this regard was rather unfortunate, since its actual value has not as yet been satisfactorily defined observationally, something the HST was designed to do. The Hubble constant (H) is defined by the equation $H_o = V/D$, where (V) is the recession velocity of a galaxy and (D) its distance from us, expressed in megaparsecs (Mpc) and in km/s, respectively. For example, the Virgo cluster of galaxies, located a little more than 50 million light-years away, is receding from us at an average speed of 1,200 km/s, while the Coma cluster at more than 300 million light-years from us, is receding at 7,000 km/s. Consequently, for an accurate determination of H, both the recession velocities and distances of galaxies must be known. The former is relatively easy to measure by spectroscopy, but accurate distance measurements are more complicated. Astronomers generally rely on several "standard candles" for this, including giant variable stars like Cepheid variables, as well as planetary nebulae, novae and supernovae. Since the absolute magnitudes of such objects are well known, they can be used as distance indicators once their apparent magnitude in other galaxies is determined.

Prior to HST and its decade-long investigation of Cepheid variables, planetary nebulae, supernovae and other standard candles in both near and distant galaxies, estimates of the Hubble constant differed by a factor of two. Depending on the kind of evidence used to calculate it, some researchers favored a H_o around 50 km/s/Mpc and others closer to 100 km/s/Mpc, meaning the age of the Universe fell somewhere between 10 billion and 20 billion years. This value is now known to be around 75 km/s/Mpc with a 10 percent margin of error. This makes the Universe about 15 billion years old, close to the estimated age of the oldest stars in it.

In addition to providing accurate data on the age of the Universe, the HST has also provided definitive information on its evolution. Two extremely long exposures known as the Hubble Deep Fields (HDF) were taken, one in a northern area of sky in the Big Dipper, the other in the southern constellation of Tucana. Both 130-hour exposures were taken with the WFPC2

camera and four filters covering the ultraviolet, visible and infrared ranges of the spectrum at wavelengths of 300, 450, 606 and 814 nm. Though the camera's field of view is only one-thirtieth that of the full Moon, almost 3,000 individual galaxies were recorded, or about two million galaxies per square degree. Expanding this into a hypothetical full-sky HST survey means that more than 100 billion galaxies would be detected at this level of sensitivity! Incidentally, such a survey would also take more than 500,000 years of continuous imaging.

Hubble also used the STIS spectrograph-imager for a 44-hour integrated light exposure (from 350 to 950 nm) of the Hubble Deep Field South region. This was the deepest and most precise probe of the Universe ever taken up to that time. It recorded objects close to magnitude 30.

The Hubble deep-field surveys have provided us with a glimpse back in space and time when the Universe was less than a billion years old. Those objects in the survey for which spectra have been obtained with the giant Keck telescopes exhibit redshift in the range of 5 to 6, but there are many more galaxies too faint to record usable spectra. Astronomers have nonetheless concluded, based on color and shapes, that many more smaller and brighter galaxies populated the Universe 12 billion years ago than do at present. Large elliptical and spiral galaxies like the Milky Way most likely formed later through collisions and mergers of the smaller galaxies in the early Universe. The truly primordial galaxies probably formed only a few hundred million years after the big bang. Objects like those, with redshifts way beyond 6, are just beginning to be detected now by Hubble and the most powerful telescopes in the world today.

A new COS spectrograph and the more advanced WFPC3 imager are slated to be installed during the next HST service mission. NASA and ESA managers have also decided that they will not service the venerable telescope indefinitely, particularly as larger, more powerful ground-based telescopes are brought into service with advanced technologies and adaptive optics. After some 20 years of outstanding service, Hubble will be decommissioned around 2009 or 2010, in favor of the Next Generation Space Telescope (NGST). ■

The SOHO Solar Observatory

I t's clear to anyone who has cast more than a passing glace at the Sun without protection that its heat would quickly damage your eyes. Remembering that those rays have already been partially filtered by the Earth's atmosphere, just imagine their intensity in space. And yet observing the Sun from space is precisely what SOHO, the joint European–U.S. orbiting observatory, has been doing since 1995. This technological marvel, sponsored by the European Space Agency (ESA) and NASA, was developed to monitor our star continuously in order to help us understand it more fully. From its 1.5 million km orbit, SOHO is a flying laboratory, equipped with instruments that can study every aspect of the physics, composition and behavior of the Sun. In so doing it provides invaluable information not just about our star but also about other stars in the cosmos.

Although we know more about the Sun than we do about any other star, our knowledge is far from complete. Even after decades of observing and modeling the Sun, many fundamental questions remain regarding its physics, internal workings and long-term behavior. Take the corona, for example. Astronomers know that it is extremely hot and visible only during total eclipses or with special filters and other instruments. Yet many questions remain. What are the physical conditions that give rise to the corona, why is it so extremely hot (several million degrees) and what fuels this process?

THE SUN AROUND THE CLOCK

The Sun emits energized, charged particles or plasma known as the solar wind, which interacts with other members of the Solar System, including Earth. It is still not clear how the solar wind is generated, what drives it and what its effects might be on the interplanetary environment.

EUROPE –
UNITED STATES

■ *SOHO was launched into a 1.5 million km high orbit above Earth to allow its battery of instruments unrestricted access to the Sun around the clock.*

Solar astronomers have a broad understanding of the Sun's workings and interior structure. Its core or nucleus is very dense with temperatures around 15 million degrees, sufficiently high to fuse hydrogen into helium. Outside the core, in the "radiation zone," energy released through nuclear fusion is converted into lower-energy X-rays and gamma rays. In the "convection zone" above that, the two million degree heat energy is gradually cooled through expansion. Temperatures are about 5,700°C by the time the photosphere is reached. The photosphere itself, the visible surface of the Sun, is a 500 km thick layer that radiates light into space. The photosphere is not smooth, but frequently marked by sunspots. These are temporary, relatively cool patches, which appear darker than the bright photosphere. Generated by the Sun's magnetic field, sunspots vary in frequency in cycles lasting about 11 years, when the solar magnetic field is reversed. The precise mechanisms underlying these cycles and the accompanying physical changes in the Sun are still not fully understood.

SOHO was launched from Cape Canaveral on December 2, 1995, to address these and other important questions about our star and stars in general. An Atlas rocket placed the satellite into a very specific orbital location, the L_1 Lagrangian point, about 1.5 million km from Earth. Named after the 18th-century French mathematician, Joseph-Louis Lagrange, this is one of five stable points in the Earth's orbital plane. Lagrange showed that when two bodies, in this case the Earth and Sun, interact gravitationally, there are five points where the forces between them are neutralized. Two of these are 1.5 million km from Earth. L_1 lies between us and the Sun and L_2 on the side of Earth opposite the Sun. L_3 and L_4 are orbital points forming equilateral triangles with point

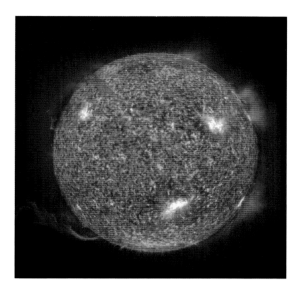

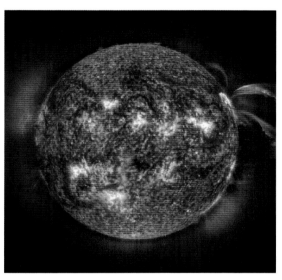

■ *SOHO is a kind of "weather" satellite that constantly monitors solar activity. These far-ultraviolet light images (30.4 nm) show the chromosphere with flares and prominences above it at temperatures around 60,000 K.*

L5 behind the Sun. By placing SOHO at point L1, it became a stationary satellite of both the Earth and the Sun, allowing it to monitor our star around the clock, something not possible with regular terrestrial satellites.

The SOHO spacecraft contains two modules: a service component for logistics purposes like energy regulation, navigation, tracking and communication, and a science module that weights 600 kg, about one-third of the observatory's total mass. The observatory measures 3.65 x 3.65 m, with solar panels extending over 9 m when fully deployed. SOHO is part of a larger collaborative program (Cluster) between ESA and NASA that is investigating the Sun and its effects on the Earth. SOHO and the dozen instruments aboard it were manufactured in Europe and launched by NASA. NASA is also in charge of control and communications through its Deep Space Network, an array of radio telescopes around the globe. Headquarters are at the Goddard Space Flight Center in Maryland. More than 500 researchers from 20 different countries are directly involved in this extraordinary project.

Three of SOHO's 12 instruments are designed to study the solar interior, which obviously can only be observed indirectly. Three special sensors, GOLF, VIRGO and SOI/MDI, monitor fluctuations in temperature, pressure and motion, all indicative of change in the Sun's interior. Five instruments, SUMER, CDS, EIT, UVCS and LASCO, target various layers of the Sun's atmosphere, particularly the inner and outer regions of the corona. The corona is a huge expanse of hot, ionized gas barely visible at optical wavelengths but a brilliant emitter of ultraviolet light and X-rays. For this reason only LASCO operates in visible light, while the other four imager-spectrographs cover the ultraviolet range and X-rays. The remaining four instruments, CELIAS, COSTEP, ERNE and SWAN investigate the solar wind. The first three sensors measure the charge density and isotope composition of energized particles in the solar wind, including electrons, protons and helium nuclei or "Alpha particles." Similar techniques are used in high-energy physics and include electrostatic sensors to measure the velocity of these particles in space. The primary function of the SWAN sensor is to map the overall shape and direction of the solar wind using the Lyman-Alpha spectral lines of hydrogen. Finally, solar ultraviolet emissions are monitored around the clock, both to establish a reference base and look for any long-term fluctuations or changes in output.

One very important element of the SOHO mission is to provide more information on any variations in solar activity. Since many aspects of this are still quite unpredictable, the various instruments aboard SOHO were designed to monitor key solar parameters continuously. This way any long-term trends as well as sudden changes in solar activity can be better correlated and better defined in terms of those parameters.

In spite of some serious technical problems in 1998, when contact with SOHO was lost for several months, the mission has worked remarkably well and all of its primary science goals have been met. Its most notable findings include: a magnetic carpet extending through the photosphere caused in part by extensive subsurface turbulence; tornado-like ejections of coronal plasma; and detailed information about the elemental composition and physical nature of the corona and solar wind. In addition SOHO has provided new information on the frequency of comets falling into the Sun and other aspects of the interplanetary environment. In short, in its decade-long existence, SOHO has provided some of the most important and fundamental information about the Sun to date. ■

The Newton X-Ray MM Space Telescope

The Newton X-Ray Multi-Mirror Space Telescope was launched from French Guyana at 15:32 universal time on December 10, 1999, by an Ariane 5 rocket. A few hours later, having reached its prescribed orbit, the satellite deployed its 16 m long solar panels. The observatory's orbit is highly elliptical, extending nearly a third of the way to the Moon at its furthest point (120,000 km) and coming to within 7,000 km from the Earth at perigee. This orbit was chosen to extend the satellite at least partially beyond the Earth's magnetosphere, that invisible shell of radiation that envelops the planet. Extending to about 120,000 km, the magnetosphere can impact satellite communication systems and also disrupt detection of high-energy radiation by orbiting observatories. Thanks to this highly elliptical orbit, the spacecraft lies beyond the magnetosphere for about 40 hours a week, during which time it carries out the bulk of its observations.

 Ariane 5, the ESA's most powerful launch vehicle, lifted the Newton X-Ray Multi-Mirror Space Telescope into an orbit that extends as far as 114,000 km from Earth. The ESA expects the observatory to function until 2010.

More than 10 m long and weighing 3.8 metric tons, Newton is by far the largest scientific satellite ever built by the European Space Agency (ESA) and required its most powerful rocket launcher. The total cost of the project in 2001 had reached about US $867 million. This included satellite construction, technical supervision and costs associated with the Ariane 5 launch. Though initially planned for only two years, the mission was subsequently extended to 10 years by ESA, since the observatory was functioning extremely well and in a near-perfect orbit. Engineers now expect it to remain in operation until 2010.

The Newton X-ray telescope, without question one of the most sensitive high-energy observatories ever built, actually combines several different types of telescopes. Conventional mirrors and telescopes that are used for ultraviolet, visible and infrared light observations will not work with the much shorter wavelength X-rays, whose much higher energy (1 to 100 nm) cannot be efficiently focused by ordinary mirrors. Instead Newton was designed with three separate "optical modules," consisting of 58 cylindrical mirrors housed one inside the other, much like Russian dolls. The mirrors are made of nickel and are coated with gold. X-rays enter one end of the cylinder and are reflected back and forth several times before being focused into CCD cameras and spectrographs. The X-ray capture efficiency of this telescope is about 60 percent. In addition to three X-ray telescopes, the observatory also carries a .3 m optical telescope sensitive in ultraviolet and visible light from 170 to 600 nm.

Depending on the wavelength utilized, Newton's CCD cameras can resolve detail between 5 and 15 arcsec. Resolution was sacrificed somewhat with the X-ray telescopes in favor of speed and high sensitivity. They can detect intensity fluctuations of less than one one-thousandth of a second, and are well suited for studies of violent events like interactions between binary stars and black holes. In fact, Newton has observed several events of this sort, including LMC X-3, which was an X-ray burst involving a binary system 170,000 light-years away in the Large Magellanic Cloud. This pair consists of a black hole and a hot blue giant orbiting each other in a mere 1.7 days. Matter from the star is periodically drawn toward the accretion disk

■ *The Lockman Deep Field in Ursa Major, as imaged by the Newton X-Ray Space Telescope at wavelengths between 1 and 100 nm. This 52-hour exposure revealed dozens of supermassive black holes hidden in the cores of galaxies and distant quasars.*

of the black hole at nearly the speed of light. This superheats the gas to several million degrees and emits an intense stream of X-rays. Although this exchange had been monitored for several years by previous satellites and remained quite steady, it suddenly flared up in 2002 just as the Newton spacecraft was undergoing routine calibration. Only about 10 minutes after this intense burst of X-rays, LMC X-3 had faded again to one one-hundredth of its peak intensity. There is as yet no satisfactory explanation for this phenomenon.

A second variable X-ray source discovered by Newton is located 2.5 million light-years away in the Andromeda Galaxy. This is binary source, XMMU J004319.4+411758, one of hundreds of X-ray sources located in various parts of this beautiful galaxy. This object is unusual in one respect, however: its emissions fluctuate with a period of only 14 minutes. Most likely this is a pair of very close dwarf stars. One is about Earth-size and the other one, just a few thousand kilometers away, is gravitationally deformed and exchanges plasma superheated to nearly one million degrees with its neighbor.

The Newton telescope also conducted a detailed, high-energy mapping survey of the Coma Cluster, a group of galaxies about 300 million light-years from us. Thanks to its exceptionally high sensitivity, the space telescope can outperform even the largest ground-based instruments at this task, and

among other findings it showed that the entire cluster is enveloped by a halo of extremely tenuous gas. Reaching temperatures in excess of 10 million degrees, this halo extends over a volume of several million light-years.

European astronomers reached a major milestone with the Newton observatory in 2000 by recording X-rays from one of the furthest objects known, the quasar J104433.04–012502 with a redshift of 5.8. An eight-hour exposure was required to register about 30 X-rays from this source.

If current cosmological models are correct, this quasar appears to us today as it was about one billion years after the big bang. The Newton data suggests that a black hole equivalent to about 3 billion solar masses lies at its core. An extremely active black hole like this is considerably larger than those found in quasars closer to us in space and time.

In early 2002 the Newton observatory also discovered a black hole located much closer to us, in the Milky Way. This 10 solar mass object, XTE J1650-50, emits X-rays generated by slowly circulating gases that are superheated to several million degrees before entering the accretion disk. This rotation appears to involve space/time itself. Predicted by Einstein, such an extreme space/time deformation is known as the Lense-Thirring effect, named after the Austrian physicist who first suggested it on relativistic grounds. The recently launched Gravity Probe B satellite will investigate this effect further. ■

■ *The Newton X-Ray Observatory, the largest satellite ever built by ESA, is over 10 m long and weighs 3.8 metric tons. The observatory is equipped with instruments to detect X-rays and ultraviolet and visible light.*

> High-energy astrophysics

High-energy astrophysics is a comparatively new field of research despite the fact that far-ultraviolet radiation, X-rays, gamma rays and cosmic rays are associated with some of the most energetic and often violent cosmic events. Techniques to investigate these processes have been advancing rapidly, however.

HIGH-ENERGY RADIATION

High-energy radiation is defined as electromagnetic wavelengths shorter than 100 nm. Radiation in the 12 to 120 electron-volt (eV) energy range, with wavelengths from 100 to 10 nm, is termed far-ultraviolet radiation, at the transition point to even more energetic X-rays. The latter have energy levels in the 120 eV to 120 thousand-volt (keV) range and wavelengths between 0.01 and 10 nm. Beyond that we find gamma rays, the shortest and most energetic waves of electromagnetic spectrum.

Although they are not true electromagnetic radiation, "cosmic rays" are even more highly energetic particles. They consist of protons, helium nuclei and nuclei of other elements, and can reach energy levels as high as several hundred gigavolts (GeV).

● The Compton Gamma Ray Observatory mapped cosmic X-ray and gamma-ray sources from space between 1991 and 2000. This 17-metric-ton satellite was one of the largest ever launched by an American space shuttle.

X-rays, gamma rays and cosmic rays are products of some of the most energetic cosmic processes and events, and typically are associated with such extreme environments as stellar coronas, supernova remnants, interacting binaries, neutron stars, black hole accretion disks and the nuclei of active galaxies and quasars.

THE BEGINNINGS OF HIGH-ENERGY ASTROPHYSICS

In addition to being highly energetic waves, X-rays and gamma rays are also difficult to detect. There are several reasons for this. First, they are completely absorbed by the Earth's atmosphere and so are only observable from high-altitude balloons, rockets or satellites. Second, these low-intensity rays are comparatively rare and not detectable by conventional telescopes.

It is not surprising therefore that effective techniques to observe X-rays and gamma rays have only been developed recently. Although X-rays from the Sun were first detected in 1948, it was not until 1962 that Scorpius X1, the first extra-solar source, was identified and not until the 1970s that the first X-ray sky surveys were undertaken. Since then, thanks to such successful space missions as EINSTEIN, EXOSAT and ROSAT, X-ray astronomy has come into its own. Today, some 40 years after the discovery of Scorpius X1, tens of thousands of additional sources have been identified.

Gamma-ray astronomy was even slower in its development. The first gamma rays located in our own galaxy were detected in the 1960s by the OSO Solar Observatory. Several so called gamma-ray "bursts" were observed

● A panoramic view of the Milky Way taken by the Chandra X-Ray Observatory shows X-ray sources at the core of our galaxy. This 120 x 48 arcmin field, from a mosaic of 30 images, covers an area of about 900 x 400 light-years. The white point sources indicate white dwarfs, pulsars and black holes. The galactic core at the center of the image shows regions of gas heated to more than 10 million degrees circling a supermassive black hole.

the following decade, but it was not until the first real gamma-ray observatory (COS-B) was launched that this area came into its own. This spacecraft mapped gamma-ray sources in the galactic plane from 1975 to 1982 and was followed by the SIGMA and Compton Gamma Ray Observatories, which discovered thousands more.

THE DETECTION OF HIGH-ENERGY RADIATION

Instruments that detect low-energy (less than a few keV) X-rays are quite unlike conventional telescopes. Since X-rays must strike detectors at very sharp angles of incidence, the ideal X-ray telescope is barrel- or tube-shaped with both flat and curved surfaces. This way, incoming photons are reflected and ricocheted onto a detecting surface or a Bragg spectrograph.

The best existing X-ray telescopes only provide about 1 arcsec resolution. They are limited not by aperture but by surface irregularities in their mirrors, even though these are polished to an accuracy of just a few atoms. X-ray interferometers are under development to overcome these limitations. The future U.S. MAXIM project, for example, proposes several X-ray mirrors in 450 km orbits that will have a combined resolving power of better than 0.01 arcsec.

The wavelength of really high-energy radiation is short enough to slip between the atoms of most solid matter. Consequently these types of photons cannot be reflected or focused in a conventional manner, but are detected through interactions with certain materials. A variety of devices have been developed that interact directly with the energy released

by gamma rays. Scintillation chambers and special semiconducting materials can count gamma-ray energy levels in the 5 million to 10 million volt (MeV) range.

What sort of resolving power do devices like this have? This depends on the wavelengths of the incoming photons. With special aperture masks, detectors like INTEGRAL can attain angular resolution of less than 1 arcmin. This device combines separate images taken through filters in different orientations into a single final image. Scintillation chambers used for even higher energy radiation can pinpoint incoming photons to within a degree, which is still considerably poorer than that obtainable with longer wavelengths. ■

● The famous galaxy M51 in Canes Venatici is shown here in an 11-hour exposure by the Chandra X-Ray Observatory. Supernova remnants and black holes appear scattered across its spiral arms.

The Chandra X-Ray Telescope

To the five astronauts aboard the space shuttle *Columbia* it was a truly beautiful sight. On July 23, 1999, the ship's cargo bay slowly opened, all moorings were released, and the Chandra satellite was slowly cast adrift against the backdrop below of the Pacific Ocean. Like a scene from a science-fiction movie, once Chandra was sufficiently far from the shuttle, powerful rocket engines lifted it into a final orbit several thousand kilometers higher.

The Chandra X-Ray Telescope is one of the most powerful space telescopes ever built. This 14 m long NASA instrument, 20 m with fully extended solar panels, weighs almost 5 metric tons. It's a technological marvel that cost more than a billion dollars, roughly twice as much as its European cousin, the Newton X-Ray Telescope.

Like the Newton, Chandra is expected to remain in operation until 2010. Just a few days after its release from the space shuttle, Chandra reached its highly elliptical working orbit of 10,000 km perigee and 140,000 km apogee, over a third of the way to the Moon. Chandra lies outside the Van Allen belt; the densest regions of the Earth's magnetosphere, for more than 55 hours of its 64-hour orbit, or about 85 percent of usable observing time.

The observatory's namesake is the renowned Indian-American astrophysicist, Subrahmanyan Chandrasekhar (1910–1995), an expert in white dwarfs, neutron stars and black holes, which are all very powerful X-ray emitters. Chandrasekhar received the

■ *X-rays readily penetrate ordinary optical mirrors. Consequently, the Chandra telescope uses cylindrically shaped mirrors made of highly reflective ceramic glass, which deflect and record X-rays internally.*

Nobel Prize for physics in 1983 for his work on the evolution of stars.

The Chandra X-Ray Telescope is sensitive to wavelengths between 0.1 and 10 nm, well below the 400 to 700 nm visible range. Because of their short wavelengths and high energy, X-rays pass right through ordinary telescope mirrors and so cannot be collected and focused in this manner. Instead, incoming X-rays must be intercepted at very low angles to ricochet like pebbles on the surface of water. Chandra is equipped with four near-cylindrical mirrors for this, to deflect the rays toward a series of cameras and spectrographs.

These mirrors are true engineering marvels. Like normal telescope mirrors, they are made of a ceramic glass called Zerodur, but in addition are polished to an accuracy of better than 1 nm and coated with iridium to make them as reflective as possible. This ultraprecise, extremely shiny surface reflects more than 90 percent of the longer wavelength X-rays, those bordering on the ultraviolet, but only 10 percent of the extremely energetic radiation.

Chandra is equipped with spectrometers and a High Resolution Camera (HRC). The latter is about 10 by .1 m in size and contains a matrix of microchannels with nearly 69 million tiny glass tubes, each 1.2 mm long but only 10 μm in diameter. When X-rays strike these lead-lined tubes, high-speed electrons are released and recorded by the camera.

As its name indicates, the HRC can resolve detail to 0.5 arcsec, an astonishing feat for the high-energy wavelengths

■ The Chandra X-Ray Telescope is sensitive to wavelengths between 0.1 and 10 nm. With 0.5 arcsec resolution, this instrument produces X-ray images comparable to those attainable with optical telescopes.

involved. In short, Chandra produces results on par with those attained by optical telescopes and the best radio telescopes.

Since early 2000, Chandra has obtained numerous spectra and images of the most violent and energetic events and objects in the Universe, including pulsars, supernovae and quasars. In August 2001, it recorded a powerful X-ray burst from the supermassive black hole at the center of our galaxy. Caused by a jet of matter heated to several million degrees, an event like this had been awaited by astronomers for nearly 20 years.

Like the Hubble Space Telescope, the Chandra spacecraft has taken

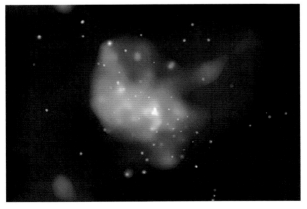

■ Invisible at optical wavelengths, the distant center of our galaxy, 28,000 light-years away, is shown here in a Chandra X-Ray Telescope image. A massive black hole is located within the very bright central region in this picture.

two deep field surveys, the Chandra Deep Field South, and another in the same region of Ursa Major as the Hubble Deep Field North. Each Chandra survey required cumulative exposure times of a million seconds (about 12 days!) and took several months to complete.

Like prior optical deep field surveys, the X-ray telescope reached back 12 billion years to the earliest stages of the Universe. The X-ray data indicated that massive black holes were far more active at that time, when the earliest galaxies were forming and dazzlingly bright stars illuminated the chaotic young cosmos. ■

The Wilkinson Microwave Anisotropy Probe

UNITED STATES

A very long voyage awaited the Wilkinson Microwave Anisotropy Probe (WMAP) the morning of June 30, 2001. Launched by a Delta II rocket, the 830 kg satellite first circled the Earth several times in highly elliptical orbits before entering a trajectory that would take it to within 1,000 km of the lunar surface a month later. The Moon's gravitational boost in turn launched the spacecraft toward Lagrangian point L_2, a stable orbital position between the Earth and the Sun, 1.5 million km distant. This is an ideal location for astronomical purposes, since the Earth, Moon and Sun all appear clustered in one region of sky, leaving the rest available for uninterrupted observing. The WMAP satellite reached that point October 1, 2001.

The primary mission of WMAP is to map the Universe as it appeared shortly after the big bang, close to 15 billion years ago. In the first few thousand years after its birth, the Universe was an extremely hot cauldron or "fluid" of energy and subatomic particles. With temperatures in the tens of thousands of degrees, this plasma consisted primarily of protons, electrons and photons. Had anyone been around to witness it, everything would have appeared like an immense fireball as bright as the Sun. With continuing expansion and inflation, the cosmos gradually cooled to about 3,000 K by the time it was 300,000 years old, cool enough for protons and electrons to combine into stable hydrogen atoms. With the conversion of plasma into atoms, the young Universe

■ *The WMAP spacecraft was placed in orbit on June 30, 2001, by a Delta II rocket. It took an additional three months before it reached its final orbit, 1.5 million km from Earth.*

became transparent and radiation was free to travel though it. These free photons are the signatures of the earliest events in the Universe. Known as the cosmic microwave background (CMB) radiation, these signals were discovered in 1965 by Americans Arno Penzias and Robert Wilson, and constituted the first direct evidence of the big bang theory.

The intense, 0.5 μm background radiation that pervaded the young Universe weakened with continued expansion and cooling, and is now detected as much longer wavelength radiation in the millimeter range. With redshifts close to 1,000, this is the oldest and most distant electromagnetic signal we can observe at present. To reach back even farther and closer still to the big bang will require entirely new types of detectors such as neutrino telescopes and gravitational wave detectors.

As a relic of the big bang, the CMB radiation is key to understanding what conditions were like in the primordial Universe and what might have given rise to its larger structure. By carefully mapping this background radiation, WMAP was looking for slight temperature variations that subsequently led to gravitational inhomogeneities and structure in the Universe. This information would also help to accurately define the values of the universal density parameter ω and the cosmological constant λ.

At first glance, the Universe appeared remarkably uniform 300,000 years after the big bang. In 1992 the Cosmic Background

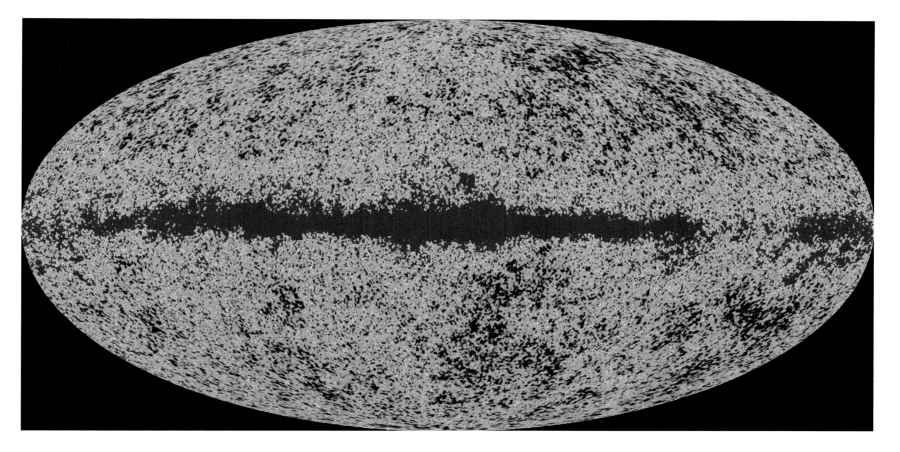

■ *This is the first detailed, all-sky picture of the infant universe. The WMAP image reveals 13-billion-year-old temperature fluctuations (shown as color differences) that correspond to the seeds that grew to become the galaxies.*

Explorer (COBE) detected only minuscule variations across an extremely cold temperature background of 2.7 K. Fluctuations were only about one part in 100,000, with 2.7251 K for the warmest regions and 2.7249 K for the coldest. The WMAP spacecraft was designed to work within that 0.0002 degree range.

A THREE-YEAR MISSION

The WMAP is equipped with two 1.5 m telescopes mounted back to back which can cover two regions of the sky 140° apart. Each telescope carries a radiometer sensitive to microwave radiation at five wavelengths: 3.3, 5, 7.5, 10 and 13.6 mm, with spatial resolution of about 20 arcmin. These instruments can measure temperatures close to 0.000,02 K. These temperature measurements are not absolute but comparative. Since WMAP takes simultaneous readings from opposite regions of the sky, it can measure even minuscule temperature differences with great accuracy. The spacecraft began its systematic survey in 2001 and covers the entire sky

■ *The WMAP spacecraft weighs 830 kg. Its two 1.5 m aperture telescopes can record millimeter wavelength radiation from two directions. NASA scientists expect this satellite to remain operational through 2006.*

every six months. Daily transmission of the results back to Earth takes about 15 minutes. Since WMAP is located quite far from us, NASA's 70 m Goldstone antenna is used to ensure that precious data are received without difficulty. The mission is scheduled to finish in 2006. Since full sky coverage is obtained every six months, the final product will be compiled from several sets of data and provide a highly accurate map of the cosmic microwave background radiation.

As soon as the first WMAP survey was completed in 2002, the results were compared to those obtained with the ground-based Cosmic Background Imager (CBI). This instrument, located at 5,100 m atop Cerro Chajnantor in the Andes, is completing a survey of cosmic background radiation in the 8 to 12 nm range. Initial results support the notion of a perfectly flat Universe. This is in accord with models that require a period of rapid inflation during which the young universe expanded even more quickly than predicted by the big bang theory. ■

The Far Ultraviolet Space Explorer

On a beautiful summer day, June 24, 1999, the Far Ultraviolet Space Explorer (FUSE) was launched from Cape Canaveral, Florida, aboard a Delta II rocket. Its mission? To explore the Universe in the far-ultraviolet range of the spectrum and hopefully answer some very fundamental questions about its chemical evolution and composition. In short, how soon were many of the chemical elements formed in the cosmos, how were they destroyed and how were they dispersed? To provide some answers to these and related questions, NASA, Johns-Hopkins University, the CNES in France and the Canadian Space Agency designed and launched FUSE. The spacecraft would study stars and the interstellar medium, planetary nebulae, galaxies and quasars in its investigations.

For nearly three centuries, telescopic observations of astronomical objects were confined to the visible portions of the spectrum. Chemists and physicists, however, were acutely aware that much information about atoms and molecules could only be attained through analysis at other wavelengths, particularly in the ultraviolet range. Obviously this was also true for astrophysics, as a full picture of our cosmos would only emerge after combining information obtained at infrared, visible and ultraviolet wavelengths.

The ultraviolet domain extends from 400 to 10 nm, and includes near-ultraviolet (above 200 nm) and far-ultraviolet wavelengths below that. Early space missions like Copernicus explored the latter domain in the 1970s and 80s, and found several unique spectral signatures between 90 and 120 nm, including those of deuterium, an isotope of hydrogen produced during the earliest stages of the Universe.

UNITED STATES – FRANCE – CANADA

■ *Launched into an 800 km high orbit in June 1999, the FUSE spacecraft circles Earth in a little over 100 minutes. Headquarters for this mission are in Baltimore, Maryland, and data are sent to a receiving station in Puerto Rico.*

Shortly after the big bang, the extremely hot young Universe was pervaded by radiation and a plasma of electrons, protons and neutrons. Only after the temperature had dropped to a few billion degrees did the first atomic nuclei form. These included: hydrogen with a single proton, the simplest and most abundant element in the cosmos; deuterium, a heavier isotope of hydrogen with 1 proton and 1 neutron; and two additional light elements, helium and lithium 7. This was the extent of primordial nucleosynthesis at that stage.

The next major stage in cosmic evolution was the formation of the first stars. These were shortlived and led to supernova explosions that both destroyed atoms and synthesized new ones. This naturally shifted the relative abundances of the elements. Deuterium, for instance, was only synthesized initially and then destroyed in stellar cores, leading to a progressive decrease in the universal deuterium/helium ratio.

Elements like oxygen VI (a highly ionized form of oxygen) also have characteristic signatures in the far ultraviolet, which makes these ions useful tracers for extremely hot gases, such as those associated with supernova explosions and stellar winds from very young stars. Molecular hydrogen, H_2, the main constituent of cold gas clouds and the interstellar medium, is another strong absorber of far-ultraviolet radiation.

Although astronomers have long appreciated the importance of the far-ultraviolet end of the spectrum, observing at those wavelengths was not possible prior to space telescopes and appropriate detectors. Conventional optical equipment, as used to observe visible and near-ultraviolet light, is simply not reflective enough for these shorter wavelengths. And since the lines of deuterium

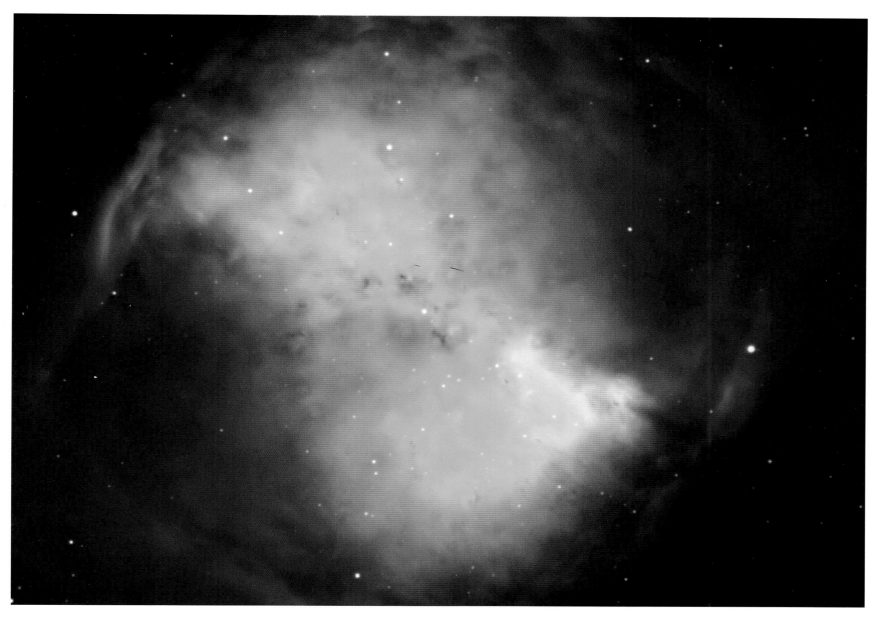

■ *This photograph of the planetary nebula M27, about 1,200 light-years from us, was taken with the one of the European VLT telescopes. The white dwarf at the center of the nebula emits most of its energy as ultraviolet radiation, turning the surrounding gas into highly ionized plasma.*

and hydrogen are relatively close together, really high-resolution spectra are required to distinguish them. The Far Ultraviolet Space Explorer was specifically designed to meet these requirements.

Though a complex and very specialized astronomical instrument, FUSE is both relatively light in weight (only 1,300 kg, compared to nearly 12 metric tons for Hubble) and inexpensive, costing only $250 million compared to billions of dollars for many other space observatories. Instead of a large primary mirror, the 5 x 1.2 x 1.8 m satellite was equipped with four .35 x .39 m mirrors, four holographic diffraction gratings and two photon counters. Redundancy was built into the system for several good reasons. To provide full coverage of the ultraviolet spectrum, specially coated optics were required: silicon carbide for 90 to 110 nm, and aluminum and lithium fluoride for the 100 to 120 nm range. Two mirrors and two holographic arrays were included for each domain to increase the instruments' overall sensitivity.

The Lyman FUSE spectrograph is the pivotal instrument aboard the spacecraft. The spectrograph's 27 x 27 cm holographic diffraction gratings, with better than 5,000 lines per millimeter, were manufactured by the French firm of Jobin-Yvon. With spectral resolution of 30,000, this outstanding instrument has provided high-resolution spectra of hundreds of stars and galaxies. Lastly, the Canadian-built Fine Error Sensor (FES) camera assures tracking accuracy to within 0.5 arcsec.

Among several discoveries, FUSE provided direct support for the existence of other planetary systems like ours, by showing that millions of comets are present in and consumed by the protoplanetary disk of the young star Beta Pictoris. FUSE also monitored ultraviolet radiation from the planetary nebula M27, whose gaseous envelope was ejected by the red giant originally at its core. This star subsequently evolved into a white dwarf with a surface temperature in excess of 130,000 K, compared to 5,200 K for the Sun's surface. This star emits most of its energy as ultraviolet radiation that causes the surrounding gas to become highly ionized. The FUSE spectra also showed that this dwarf is circled by a cloud of molecular hydrogen (H_2) that is heated to more than 2,500 K. It is unclear how these molecules are shielded from the intense temperatures and why they are not ionized like most of the interstellar medium. Clearly, answers to these questions must come with the next generation of space telescopes. ■

> The many crafts of astronomy

The scope of modern astrophysics is nothing less than a complete account of the birth, structure and evolution of celestial objects and the cosmos as a whole. Unlike classical astronomy, which was concerned primarily with planetary motions and celestial mechanics, modern astrophysics does not draw only on mathematics, but on many other branches of science as well.

AT THE CROSSROADS OF SCIENCE

Astronomy is branching more and more into other areas of science. Basic physics is at its core of course, including optics, mechanics, electricity and magnetism, nuclear physics, solid state physics, gravitational and relativistic physics, geophysics, and so forth. The area known as "applied" physics is key in such areas as telescope design, instrumentation and related engineering. Advanced mathematics and computer science are vital in the areas of theoretical astronomy, including modeling and simulation, and chemical physics applies to the understanding of the chemistry of stellar and planetary atmospheres, the interstellar medium, etc. Finally, more and more links are being established with biology, particularly with respect to issues of the origin and evolution of life.

Because of the extraordinary complexity of modern astrophysics, specialization into subdisciplines like solar and planetary astronomy, stellar and galactic astronomy, and cosmology is inevitable. Clearly, however, these fields are all interconnected and researchers must stay broadly up to date in most of them in addition to their own specialization. This is crucial, since advances in one area can directly impact another, and provide essential information for theorist and model builders, or for the development of new technical approaches and instrumentation.

Increasingly, progress in astronomy is driven by advances in observational methods and technologies, including the development of sophisticated new instruments, innovative telescope designs, space-based observatories, and so forth. In addition to highly specialized engineers and technicians, this also involves specialists in optics, electronics, mechanics, computational sciences and many other fields.

EDUCATIONAL REQUIREMENTS

Most astronomers have a solid background in physics or engineering physics. Depending on the state or country in which they are educated, they may acquire this as undergraduates at polytechnical colleges or universities. This may be followed

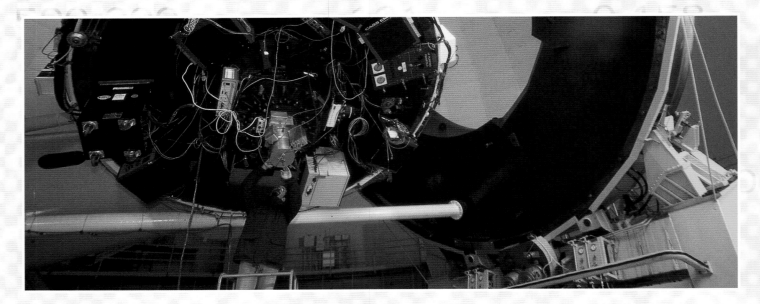

● Expertise in many areas is needed to ensure the flawless function of a 300-metric-ton telescope that is equipped with optics accurate to within a few nanometers and batteries of highly sensitive receivers. Since the cost of operating a modern, giant telescope is close to $10,000 an hour, astronomers, technicians and engineers work closely together to ensure smooth operation of the facility.

by four years postgraduate work at the master's level. After that students typically enter doctoral studies at a university, which will involve a research dissertation or thesis in an area of astrophysics or a related field. Research may be carried out at their home university, but might also involve work at a foreign institution or abroad. Once they have earned the doctoral degree, most young researchers continue as postdoctoral scientists abroad. This provides them with additional research and work experience and broadens their horizons.

Engineering students typically attend either university or specialized engineering colleges. Although they generally require less formal study than research scientists, engineers still need at least five years to earn a degree in their specialized field.

Technicians and technical support personnel typically require a solid undergraduate degree in a basic science, plus a couple of additional years beyond that in areas like physics, computer science, electronics, etc.

THE DISSEMINATION OF KNOWLEDGE

An important aspect of scientific research is the dissemination of results through publication in dedicated journals and through presentations at national and international conferences. Such exchanges are crucial not only for the exchange of information and ideas, but also to establish new collaborations and projects.

Teaching and mentoring young scientists is also something most researchers take very seriously. This is particularly important at the university level, where students acquire observation and research experience through short-term projects and courses in different areas of astronomy.

Researchers also have the responsibility of communicating their findings and discoveries to non-scientists and to the public. Many do this through lectures at schools and other organizations, through open-house events, popular magazines and other media. Astronomers have plenty of opportunities to inform others about their latest findings, while also educating the public about the methods and practices of their chosen profession. Most importantly, they can share their enthusiasm about subjects that never cease to intrigue human beings.

THE ORGANIZATION OF ASTRONOMY IN FRANCE

In France, astronomical research is funded by several major organizations, including the National Council for Scientific

● This is the control room of the 3.6 m Canada–France–Hawaii Telescope. While technicians monitor all instrument functions and atmospheric conditions, astronomers concentrate on observing and collecting data.

Research (CNRS) and various universities and astronomers' professional associations (these latter two are under the aegis of the Ministry of Higher Education). The Atomic Energy Commission (CEA) also supports some research projects. In addition, several collaborative projects exist between the aforementioned organizations and the French space agency, the National Center for Space Studies (CNES).

Like other complex areas of science, astronomy and astrophysics depend on often costly facilities and infrastructure, and much of this is available only through multinational support and collaboration. In space, for example, France has collaborated with other nations through the European Space Agency (ESA), as well as with NASA and the Russian Space Agency, especially in launching many outstanding space telescopes and other instruments. Collaboration has also been key in the development and funding of ground-based facilities. The European Southern Observatory (ESO), with which France has been associated since its formation in 1962, represents the first major joint European effort in ground-based astronomy. Since then, similar collaborations have followed, including the Thémis Solar Observatory on Pico de Teide in Tenerife, a Franco-Italian project, and outside Europe, the Canada–France–Hawaii Telescope (CFHT). For the last 25 years, the latter has provided the French astronomical community access to one of the world's finest 4 m class telescopes. Finally, due to its sheer size, the giant ALMA project, which is slated to begin in 2010, represents a collaboration of truly global proportions. ∎

The Spitzer Space Telescope

aunched from Cape Canaveral aboard a Delta II rocket in August 2003, the Spitzer Space Telescope (SST) is the last of NASA's four great observatories. Originally called the Space Infrared Telescope Facility (SIRTF), the telescope was subsequently renamed in honor of the late Lyman Spitzer, the astronomer who first suggested placing a telescope in space. This $765-million, 1-metric-ton telescope, equipped with 0.85 m Ritchie-Chrétien optics, will observe in the infrared at wavelengths between 3 and 180 μm. As a result, coverage of the electromagnetic spectrum now extends from 0.1 to 10 nm with the Chandra X-Ray Telescope, through 0.3 to 2.5 μm with the Hubble Space Telescope, and now into the infrared with Spitzer.

The powerful Delta II rocket placed the SST into an Earth-trailing solar orbit about 15 million km from home. This was done for several important reasons. First, this removed the telescope as much as possible from any interfering thermal radiation it would encounter from the Earth, Moon and Sun in a terrestrial orbit. Second, by placing the SST into "deep space," the telescope is guaranteed an extremely cold environment, ideal for infrared astronomy. At that location, the ambient temperature is around 40 K, or –233°C. This natural cooling saves on the cryogenic liquid helium needed to operate the three principal instruments on board: the IRAC camera, the IRS spectrometer and the MIPS photometer, all of which must be cooled to around 2 K, or –271°C. With the low ambient temperature and a supply of 360 liters of liquid helium, the telescope is expected to keep operating well into 2008.

The telescope's f/12 beryllium mirror, with 10 m focal length, weighs less than 50 kg. The focal-plane IRAC camera contains four infrared detectors that function concurrently. Each 256 x 256 pixel detector covers a 5 x 5 arcmin field of view. The array can capture images at wavelengths of 3.6, 4.5, 5.8 and 8 μm, with about 2 arcsec resolution. This may seem low compared to 0.5 arcsec for the Chandra space telescope and 0.1 arcsec for Hubble, but it is remarkably good given Spitzer's modest aperture and the long wavelengths involved. Thanks to its optimal location in space, very sensitive receptors and wide field of view, the SST has already produced some truly remarkable infrared images.

The universe studied by the SST is essentially a very cold universe, which includes dark nebulae and stellar nurseries, the interstellar medium and dusty regions around young stars with new planetary systems in the making. It also includes dwarf stars and supergiants in the star-forming regions of remote galaxies.

The telescope's IRAC camera will for the first time allow us to observe deep within dusty nebulae like the Horsehead in Orion, and directly view the embryonic star systems inside. Spitzer will also investigate protoplanetary disks around young stars, which radiate strongly in the infrared. By surveying several hundred nearby stars, the telescope should provide a large enough sample to permit reliable estimates on the frequency and distribution of planetary systems in our galaxy. However, due to its limited resolving power, the SST will not be able to directly

■ The Spitzer Space Telescope (SST) is equipped with an 0.85 m beryllium mirror with a focal length of 10 m. The telescope's cameras and sensors are cooled to 2 K (–271°C) by liquid helium.

■ *The magnificent spiral galaxy M81, located some 12 million light-years from us in Ursa Major, as recorded by the multiband imaging photometer and infrared camera of the Spitzer Space Telescope. Thermal infrared emission images at 24 microns (red, left), 8 microns (green, center) and 3.6 microns (blue, right) are shown at bottom and a composite mosaic of all three as the main image. The insert at top right was taken in visible light by a ground-based telescope at Kitt Peak National Observatory. The Spitzer images highlight dust particles, gas and star-forming regions in the galaxy's spiral arms.*

image planets around nearby stars; that is something only the next generation of space telescopes will be capable of.

Farther afield, the SST will provide new information on galaxy formation and evolution over cosmic time. One project, the Spitzer Infrared Nearby Galaxy Survey (SINGS), will examine star-forming rates in the relatively nearby galaxies of the Virgo cluster. Another project, the Spitzer Wide-area Infrared Extragalactic survey (SWIRE), will map galaxy distribution out to a redshift of about 2.5. This is possible because visible light (0.5 μm) from very distant objects is shifted to longer wavelengths, and can therefore be detected at 3.6 μm in the infrared. The Spitzer telescope will also reexamine the Hubble Deep Field North area in Ursa Major at various infrared wavelengths and perhaps image some of the most distant galaxies in the Universe. ■

The High Energy Transient Explorer

On the morning of October 9, 2000, under blue tropical skies, NASA's Lockheed L1011 took off from the Kwajalein atoll in the Marshall Islands. The heavy duty U.S. transport plane carried a 15 m long, 19-metric-ton Pegasus rocket under its fuselage. This small three-stage rocket was flown to an altitude of 12 km carrying a small satellite, the High Energy Transient Explorer (HETE). The spacecraft's mission is to monitor some of the most mysterious events in the Universe, powerful explosions known as gamma-ray bursts (GRB). A few minutes after its release at close to 1,000 km/h, the Pegasus engines ignited, carrying its payload toward a 600 km high orbit.

Gamma-ray bursts were first detected in 1967 by U.S. surveillance Vela satellites, which were deployed to ensure that the Soviet Union did not violate treaties to limit testing of nuclear weapons in the atmosphere. Any clandestine tests would be detected by these satellites as sudden "bursts" of gamma-ray radiation. As things turned out, much to the surprise of the U.S. military, GRBs were frequently detected, not coming from Earth but from far out in space!

UNITED STATES – EUROPE – INDIA

■ *Weighing only 120 kg, the HETE spacecraft was launched into a 600 km high orbit in 2000 to monitor hypernova explosions producing gamma rays.*

ONE HYPERNOVA A DAY

Once it became clear that these cosmic explosions were not caused by the Soviet regime, the Pentagon "declassified" them and provided astronomers access to the Vela satellite data. Scientists have been struggling to explain them ever since. The satellites registered an average of one GRB per day: intense flashes lasting from a fraction of a second to half a minute or more.

NASA launched the Compton Gamma Ray Observatory in 1991 to specifically investigate these phenomena. Over 2,000 bursts were recorded during the following years, distributed all across the sky. Only one plausible explanation seemed to fit this pattern: GRBs must be coming from very distant galaxies. If that was the case, however, these bursts must be extremely powerful explosions, more powerful even than the most violent supernova.

One proposed cause of GRBs is the collapse of closely paired neutron stars. These extremely dense, rapidly orbiting bodies progressively dissipate gravitational energy and eventually combine. The fusion of two such stars should theoretically result in a brief but tremendously powerful explosion.

One problem facing researchers trying to pinpoint the exact celestial location of individual GRBs was the comparatively poor resolving power (tens of arc minutes) of satellites monitoring gamma rays. How could such objects be differentiated against a background of tens of thousands of distant galaxies? The answer came on May 8, 1997, when the Italian-Dutch satellite Beppo SAX detected an intense 15-second GRB and provided accurate positional information on its source. This opened the way for major observatories around the world (from La Palma to the VLA to the HST) to observe it at various wavelengths, and for the Keck I telescope in Hawaii to secure a spectrum of the faint 20th magnitude object. Located in the constellation of Cameleopardalis, it was a star in a distant galaxy with a redshift of 0.83. This corresponds to a

■ *Despite their brief but intense emission of gamma rays and X-rays, gamma-ray bursts (GRB) are difficult to detect at optical wavelengths, as indicated in this image of GRB 000131, taken with the European VLT.*

distance of eight billion light-years, more than half the age of the Universe, thus supporting the theory that GRBs are rare, gigantic explosions at enormous distances from us.

We now know a lot more about gamma-ray bursts, since many more have been observed, and most exhibit very high redshifts. It is likely that they either represent fusions of neutron stars or explosions of supermassive stars followed by formation of black holes. Such events are now termed "hypernovae." It is difficult to grasp the scale of the energy involved. In January 2000, for example, GBR 000131 occurred in a galaxy in Carina with a redshift value $z = 4.5$. The energy released during this 30-second explosion was estimated around 10^{51} ergs, more than all the energy the Sun has released in its five billion years of existence!

HETE was designed to detect X-ray and gamma-rays bursts and also to provide immediate positional data to a worldwide network of scientists, allowing them to monitor the event at various wavelengths. Led by the Massachusetts Institute of Technology, the program includes several other American laboratories, the National Center for Space Studies (CNES) and the National School for Space and Higher Aeronautics (SUP'AERO) in France, the Italian National Research Council (CNR), the National Institute of Space Research (INPE) in Spain and, lastly, the Tata Institute of Fundamental Research (TIFR) in India. Initiated in early 2001, the program is expected to continue to 2005. HETE is a small, 120 kg satellite about 1 m in diameter. It can detect X-ray and gamma-ray bursts with positional accuracy of 10 arcmin and alert the worldwide monitoring network in less than a minute. Since entering service HETE has cataloged several hundred GRB.

Due to their enormous distances and extremely powerful nature, GRBs have been used as "standard candles" by astronomers for observations of the earliest stars and galaxies in the Universe. In theory the HETE spacecraft can detect GRBs with redshifts between 10 and 20, or to less than 100 million years after the big bang. ■

The INTEGRAL Space Telescope

The International Gamma Ray Astrophysics Laboratory (INTEGRAL) was successfully launched on October 17, 2002, from Baikonur in Kazakhstan. Financed jointly by the European Space Agency (ESA) and Russia, INTEGRAL's mission is to produce a complete map of the sky in the high-energy gamma-ray waveband. However, the spacecraft is also equipped with X-ray detectors and an optical camera to simultaneously observe at those wavelengths as well. The 4.1-metric-ton satellite is 5 m long and 3.7 m in diameter. For economic reasons, the ESA kept the same overall design for both INTEGRAL and the Newton X-Ray MM Space Telescope. Although the spacecraft cost only about US $433 million, it is expected to remain functional to at least 2008.

Russia provided its powerful Proton rocket to launch the spacecraft in exchange for observing time on this new European observatory. The telescope was placed into a highly elliptical orbit, 10,000 km at its closet to Earth and out to a 153,000 km apogee, more than halfway the distance to the Moon. With an orbital period of 72 hours, the spacecraft is above the magnetosphere (40,000 km) about 90 percent of the time.

Though relatively infrequent, gamma rays are the most energetic form of electromagnetic radiation. Because of this, they readily penetrate most types of matter and like high-energy X-rays, cannot be captured and focused by mirrors like other forms of radiation. Gamma rays are such energetic photons that a single one packs between a million to a billion times more energy than visible light.

EUROPE–RUSSIA

■ *Europe's INTEGRAL space telescope is similar in design to the Newton X-Ray Telescope. The 4.1-metric-ton spacecraft is 5 m long and 3.7 m in diameter, and is equipped with gamma-ray spectrometers and an imager.*

Being of such extremely short wavelengths, between 0.01 and 0.000,001 nm, they readily pass between atoms of most matter except for very dense materials. However, fortunately for life on Earth, gamma rays are almost completely absorbed by our atmosphere.

Because of this, and the difficulty of capturing them effectively, the gamma-ray "window" was the last one to be opened to astronomy; in the late 1970s the first satellite detectors were placed in orbit, including COS-B, SIGMA and the Compton gamma-ray observatories. Gamma-ray observations continued to be problematic, however. Since they come from every direction in the sky and are not easily channeled, it is difficult to determine their exact point of origin. For this reason, the INTEGRAL telescope and its detectors are heavily shielded to minimize interference by random background gamma rays. Since no effective way exists as yet to channel gamma rays, the telescope's optics are like a 19th-century camera obscura, where a pinhole-size aperture admits a pencil-thin beam of light that projects an image inside the box.

INTEGRAL's coded objective masks work in a similar manner. Made of hard, thick tungsten, each mask is perforated to admit gamma rays in one direction only and project them onto detectors at the telescope's focal plane, 3 m within the fully shielded spacecraft. Computers determine the angle of incident for each gamma ray entering the masks, to trace it back to its point of origin in the sky. Since gamma rays are so infrequent, hundreds of hours of exposure time is needed before sufficient numbers are recorded.

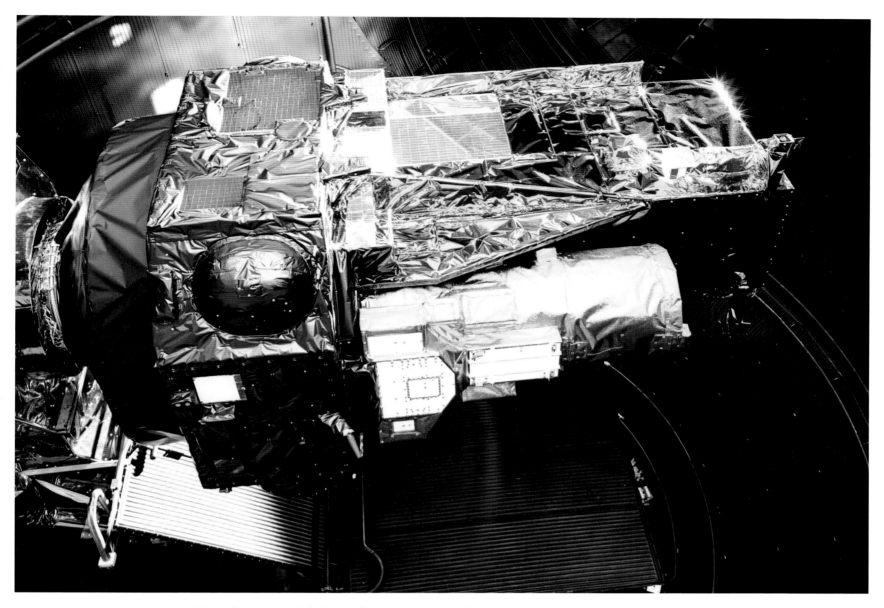

■ Prior to launch in the fall of 2002, the INTEGRAL space telescope was successfully tested in the European Space Agency's Large Space Simulator (LSS) in the Netherlands. This huge test chamber operates under high-vacuum and solar-radiation conditions similar to those encountered in deep space.

All instruments aboard INTEGRAL are co-aligned and observe the same region of sky simultaneously. To prevent extraneous gamma rays from striking it, the highly sensitive SPI spectrometer is so fully shielded that it alone weighs more than 1.3 metric tons. The IBIS imager, the best gamma-ray telescope to date, covers a field of more than 350 square degrees with 720 arcsec resolution. This is the highest level of resolution attainable at present in this high-energy domain. As a point of reference, the resolving power of the naked eye is about 60 arcsec, and the highest resolution attained to date with the most powerful optical, infrared and radio telescopes is around 0.1 arcsec. With the level of resolution attained with INTEGRAL, all celestial objects appear rather fuzzy and about half the size of the full Moon.

■ Though highly energetic, gamma rays are so infrequent that they must be recorded one at a time. The INTEGRAL space telescope will monitor gamma rays and X-rays emitted by black holes and violent supernova explosions.

In addition to the spectrometer and gamma-ray imager, INTEGRAL also carries two support instruments to help locate gamma-ray and X-ray sources in the sky, and to monitor objects that vary in energy output. This includes an X-ray monitor with 3 arcmin resolution and an optical monitoring CCD camera with a 50 mm lens capable of reaching 19.7 magnitude objects at about 1 arcsec resolution.

Gamma rays are produced in some of the most extreme physical environments known, including the cores of massive stars with temperatures exceeding several million degrees and their supernova remnants. In addition to providing more information about these most violent events in the cosmos, INTEGRAL will also study nuclear reactions on the surface of white dwarfs, neutron stars and superheated gases about to enter black holes. ■

The Einstein Gravity Probe B

Gravity Probe B (GP-B) was launched on April 20, 2004, from Vandenberg Air Force Base in California aboard a Boeing Delta II rocket. Although this joint NASA–Stanford University mission is not strictly speaking an astronomical endeavor, the probe addresses questions of fundamental importance to physics, astronomy and cosmology.

Placed in a polar orbit some 640 km above Earth, GP-B will spend about two years testing two key predictions of Einstein's general theory of relativity, first proposed in 1916. This theory is essentially a theory about gravity. Einstein proposed that energy, mass and space/time are intimately intertwined, and that space/time is distorted or warped by mass or energy. The theory of general relativity placed cosmology on solid scientific footing for the first time, and later led directly to the big bang theory of the origin of the Universe. For nearly a century since Einstein published his theory, astronomers have observed and verified many predictions stemming from it, including the expansion of the Universe, the curvature and warping of space/time, gravitational distortions of spectra, etc. Moreover, many of the predictions of general relativity have been repeatedly validated by studies of very compact objects in the cosmos, including white dwarfs, neutron stars and black holes.

TESTING GENERAL RELATIVITY

American astronomers Joseph Taylor and Russell Hulse provided strong evidence for the existence of gravitational radiation, as predicted by Einstein in 1916, through their observations of the

UNITED STATES

■ *At the core of the GP-B experiment are four fused-quartz spheres, spinning freely in a vacuum, 6,000 times a minute.*

inward spiraling binary pulsar PSR 1913+16. The two investigators interpreted the tremendous quantities of energy emitted by these objects as gravitational waves warping the space/time continuum like waves on water. Further observations of these phenomena are being undertaken by the Virgo and Ligo interferometers.

Although general relativity has passed every test and prediction since it was first proposed, scientists realize that the theory is necessarily incomplete and will eventually be replaced by a more general theory of everything, particularly in the subatomic realm. While models of general relativity work well at the cosmological level, they do not adequately deal with the physics of the extremely small. The equations of relativity define space/time clearly as a dynamic continuum encompassing the entire Universe.

The subatomic world, however, is best defined in terms of quantum mechanics, another great theory of the early 20th century. In many ways the laws of quantum mechanics are incompatible with general relativity. The latter requires that the position of particles in space/time be clearly defined, while quantum mechanics holds that the exact location and motion of subatomic particles cannot be known simultaneously. Their behavior is probabilistic or uncertain, and not predictable. Both theories work well at their respective levels in the Universe, but come into conflict in defining the space/time singularity of black holes or the moment of the big bang. If general relativity fails to explain the infinitely small, quantum mechanics effectively ignores the large-scale space/time structure of the cosmos.

This poses a fundamental dilemma for cosmologists and particle physicists alike, suggesting that either general relativity or quantum mechanics is incorrect at some level. That is why both physicists and astronomers await the results of the Einstein gravity probe with such anticipation. GP-B will measure two facets of Einstein's theory: how a massive body like Earth warps and drags the fabric of space and time around it. This effect (known as the "geodetic effect" or the Lense-Thirring effect, after the Austrian scientists who predicted it) is extremely small, so small in fact that it has not been possible so far to observe it.

To look for any such effects associated with the Earth itself, GP-B clearly requires an independent, external reference point. It will use a .14 m telescope for this, pointed permanently at HR 8703, a distant star in Pisces. In theory, the spacecraft should "twist" in space/time by 0.042 arcsec with each orbit around the Earth. To measure such a small quantity reliably requires a unique and highly imaginative approach. The test mass at the core of the experiment must float freely in space, totally isolated from external effects, such as the Earth's magnetic field, atmospheric drag and solar radiation. The satellite must constantly compensate for its own motion, yet leave the test mass free to react to any Earth-induced distortions in space and time.

At the heart of the experiment are four gyroscopes: electrically supported spheres spinning in a vacuum at 6,000 revolutions a minute. With surface accuracies of a few nanometers, the 38 mm diameter fused-quartz spheres are the most perfectly round objects ever made. In other words, if these spheres were scaled to the size of the Earth, their surfaces would be smooth to within 5 m. Except for neutron stars, no more perfect a sphere exists. The GP-B experiment is probably one of the most sensitive physics experiments

■ *The Gravity Probe B was launched to test one of the predictions of the general theory of relativity, the Lense-Thirring effect. According to Einstein's theory, space/time is warped near compact, massive bodies rotating in space.*

ever undertaken. All its key components, including the gyroscopes, test mass and tracking telescope, are constructed of fused quartz, one of the most homogeneous and stable substances known. In addition, the tiny spheres are enclosed in a metal chamber chilled to –271°C, a mere 1.8° above absolute zero.

Two of the gyroscopes spin in a clockwise direction, and two counterclockwise. The Lense-Thirring effect, which distorts space/time in the direction of the Earth's rotation, should impart a minute perpendicular torsion on the gyroscopes.

It is widely hoped by physicists and cosmologists that this experiment will not support general relativity. Though verified repeatedly through very precise measurements, Einstein's theory is nonetheless limiting. If the GP-B experiment does not fully support the predicted values of the Lense-Thirring effect, theorists will be in a position to derive equations for a more universal theory wherein general relativity is only an approximation. Theorists have been working hard for the past quarter-century to develop alternative solutions to Einstein's theory and to bridge the huge conceptual gap between the big bang theory and the recent theories describing black holes. String theory is drawing much attention, since it promises to bridge both general relativity and quantum mechanics.

Another satellite testing a fundamental principle of physics is planned for 2010. The Satellite Test of the Equivalence Principle (STEP), a joint U.S.–European mission, will test to within 10^{-18} the principle of equivalence first formulated in the 18th century by Galileo. This principle states that all bodies are equally influenced by gravity irrespective of their composition. Einstein explained this in terms of general relativity by stating that gravity is not a force but an effect due to the curvature of space, caused by mass. ■

Telescopes of the Future

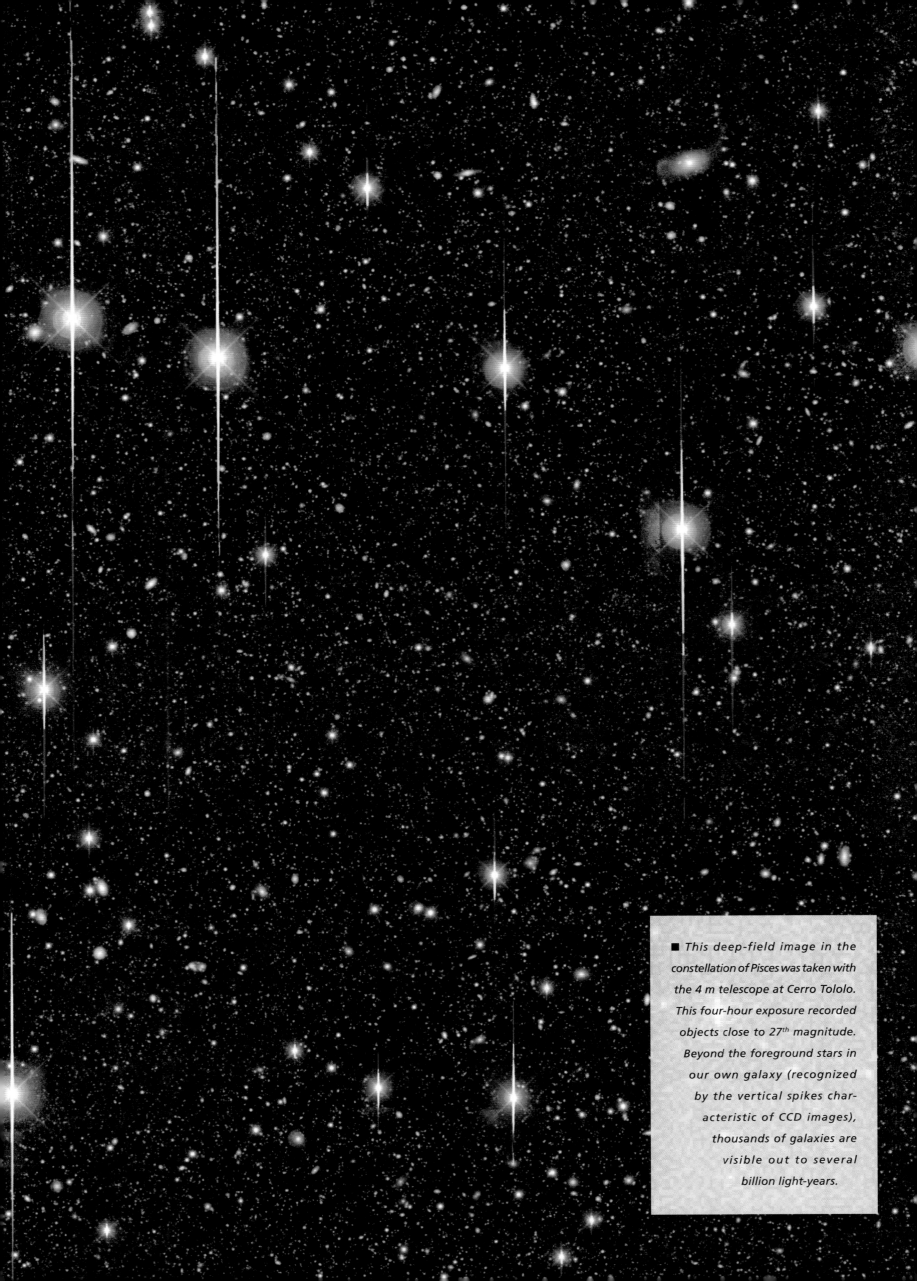

■ *This deep-field image in the constellation of Pisces was taken with the 4 m telescope at Cerro Tololo. This four-hour exposure recorded objects close to 27th magnitude. Beyond the foreground stars in our own galaxy (recognized by the vertical spikes characteristic of CCD images), thousands of galaxies are visible out to several billion light-years.*

The Atacama Large Millimeter Array

UNITED STATES –
EUROPE–JAPAN

The dusty road from the small town of San Pedro de Atacama climbs sharply eastward toward the Andean cordillera. Making its way toward Paso de Jama, the pass marking the border between Chile and Argentina, the road winds past the majestic volcano Licancabur and the glacial waters of Laguna Verde. Along the way, you come across an occasional llama, casually grazing on *ichu*, the sparse shrubbery of the altiplano. After a 30 km climb into the cold, dry air, you come to a fork in the road with a makeshift sign directing you toward the right. Seemingly leading nowhere, a trail meanders past scattered glaciers slowly sublimating in the extremely dry air. There is no sign of life here at the foot of Cerro Chajnantor — no flora, no fauna — just desert dust, lava, rocks and relentless sunshine under the deep blue sky.

■ *A truly international project, ALMA involves the six member countries of the ESO, as well as Chile, Japan, Canada and the United States.*

Only then do you notice a smattering of storage buildings across the landscape. These house computers, food and other supplies, a weather station, plenty of solar panels and satellite communication gear. Even though nothing much changes from day to day, a webcam automatically scans the scene every couple of hours and transmits the images via the Internet. Apart from occasional cirrus clouds and a dusting of snow in the winter, the sky is perennially clear.

The webcam images are reminiscent of the Viking and Pathfinder images on Mars, but then, this really is an alien world as well. At an elevation of 5,100 m, the terrain around Cerro Chajnantor is harsh and inhospitable. Barometric pressure is about half that at sea level, solar radiation is intense and dangerous, and the air is the driest on the planet. The transparency of the atmosphere, however, is without equal.

Cerro Chajnantor's potential for astronomy was first recognized in the 1990s and subsequently validated by several research groups. In the late 1990s the site was selected for one of the most ambitious astronomical facilities in the world, the Atacama Large Millimeter Array (ALMA), an international undertaking without precedent. ALMA will involve Europe, North America, Japan and Chile. By combining the European LAS interferometer project, the American MMA and Japan's LMSA (all similar in concept and design), a larger, more effective project emerged.

Slated for completion between 2005 and 2010, the project will cost approximately US $743 million. Consisting of some sixty-four 12 m antennas, with a total effective collecting area of 7,200 square meters, the array will observe at wavelengths ranging from 10 mm to 350 μm. When finished, ALMA will be one of the most sensitive astronomical facilities ever built for any wavelength.

Construction of ALMA will build on the experience gained with the Franco-German interferometer on the Bure plateau in the Alps, whose six movable 15 m dishes were an obvious precursor to the Chilean project. Like them, the ALMA antennas will be constructed of carbon fiber with highly reflective aluminum surfaces accurate to about 20 μm.

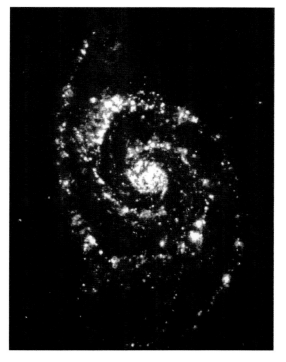

■ *This simulated image of the well known spiral galaxy M51 illustrates the 0.1 arcsec resolution expected of ALMA in the millimeter wavelength range.*

Each of the 64 steerable telescopes weighs close to 50 metric tons, and can be repositioned by special transporter into six increasingly larger configurations: 150, 325 and 680 m, and 1.4, 3 and 12 km. Laid out in two nearly concentric circles, ALMA's configuration will alternate between the inner and outer circles to minimize the number of antennas that have to be rearranged during each reconfiguration.

In its compact configurations, the interferometer will be used for real-time widefield observing, as typically done with single aperture telescopes. In its widest configurations, for really high-resolution work, the array will have to dedicate many hours of integration time per observing session.

ALMA is designed to provide data in the millimeter and submillimeter range comparable to those of the most powerful optical and infrared telescopes. For example, in its compact 150 m configuration, the array will produce images with 1 arcsec resolution at wavelengths of 1 mm. With its 1.4 km baseline, resolution near 0.1 arcsec is expected, comparable to the Hubble Space Telescope and 8 to 10 m class telescopes like Yepun, Keck and Gemini. With its maximum baseline of 12 km, ALMA is expected to reach 0.01 arcsec resolution, equivalent to that of the VLTI operating in the infrared at 10 μm. It is also anticipated that in its most compact configuration, ALMA will provide 15 x 15 arcmin fields of view, similar to those

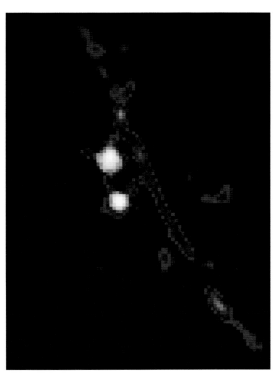

■ *This 1.2 mm image of IRAS 17175-3544 was taken with the SEST telescope at La Silla Observatory. ALMA is expected to provide a hundred times more detail.*

provided by existing giant telescopes.

The facilities control center for ALMA will be in the picturesque village of San Pedro de Atacama, located at the base of the plateau, a "mere" 2,450 m above sea level. Since it is both dangerous and uncomfortable for people to spend extended time at extreme altitudes, the ALMA technical and maintenance staff will only be allowed at Cerro Chajnantor for limited periods of time, and the few buildings at the site will be pressurized.

Although all areas of astronomy will benefit from this imposing new facility, researchers into the coldest reaches of space will use it the most. The interferometer will be able to "see" through dark clouds and nebulae, areas largely invisible at optical and infrared wavelengths. Stellar nurseries like the Orion Nebula will be scrutinized in detail, especially the protoplanetary globules and proplyds. Dust disks and embryonic planetary systems as far out as Taurus, at 450 light-years, will be studied with previously impossible angular and spectral resolution. At those distances, ALMA will be able to resolve detail in the order of 300 to 600 million km (between 2 and 4 AU), and reveal fine detail in the gas and dust disks around stars, including embryonic exoplanets leaving a trail of accreting matter. The more massive young planets may even be imaged directly. The key will be to observe planetary formation at various stages by sampling several dozen stellar dust disks.

■ *This is an artist's rendition of what the Atacama Large Millimeter Array will look like upon completion in 2010. This site at Cerro Chajnantor in the*

OBSERVING THE VERY FIRST GALAXIES

Since ALMA will provide resolution in the millimeter domain comparable to what Hubble can do in visible light, solar-system astronomers will use it to study the atmospheres of Venus and Mars for accurate temperature measurement and water-vapor content at various altitudes.

However, ALMA's most important contributions will likely be in the areas of galaxy formation and cosmology. In the coming decade, working in conjunction with the Next Generation Space Telescope (NGST), ALMA will finally provide researchers with their first glimpses into the "dark age" of the Universe, something not currently possible. In other words, they will be able to study objects with redshift greater than 6: galaxies and stars that formed during the first billion years of the Universe's existence. According to current theories, these massive objects formed in a much denser, colder environment, out of the first hydrogen and

Chilean Andes was chosen after several years of planning. At an elevation of 5,100 m this immense plateau is probably the most arid spot on Earth.

helium to condense out after the big bang. Most theorists think that the earliest stars were shortlived blue supergiants equivalent to between 100 and 200 solar masses. The forces and matter released when this generation of stars exploded at the end of their lives in turn triggered several successive rounds of star formation. Hubble and Keck I observations in 2001 of the Abell 2218 galaxy cluster indicate that the very first objects in the Universe were small star clusters just a few hundred light-years across. Populated by massive stars a million times larger than the Sun, these clusters eventually merged gravitationally to form the first galaxies.

Rich in gas and dust, these protogalaxies radiated strongly at infrared wavelengths. After some 15 billion years of cosmic expansion, these emissions have been shifted into the millimeter range. Theoretically, ALMA should be able to detect objects with redshifts between 15 and 20, or between 100 million to 300 million years after the big bang. ∎

The COROT Space Telescope

FRANCE – EUROPE

I f all goes as scheduled, in 2006 a Russian Soyuz rocket will take off from the Baikonur launch facility in Kazakhstan carrying the French COROT satellite into orbit. This modest mission, led by the National Center for Space Studies (CNES) in collaboration with the European Space Agency (ESA) and several other European partners, is dedicated to the search for Earth-like planets around other stars. If successful, this would be one of the most important discoveries ever.

Weighing only 350 kg, the COnvection ROtation and Transits (COROT) satellite cost around US $62 million and will involve a small .27 m telescope equipped with a widefield CCD sensor. With a 900 km high polar orbit, COROT will be well above the outer fringes of the atmosphere and its blurring effects. This will provide ideal conditions for high-precision stellar photometry not possible at lower elevations.

Conceived in the early 1990s, the COROT mission was primarily intended for astroseismology, a branch of research that can only be done from space. French astrophysicists plan to study the internal structure and dynamics of nearby stars by observing their natural oscillation modes.

HUNDREDS OF STELLAR ECLIPSES

Seismology has provided a great deal of information about the inner workings of the Earth, the Moon and the Sun, and should provide comparable insight into the complex reactions inside stars. COROT will attempt to do this through very high-precision stellar photometry.

Precision photometry of this type, however, can also be used

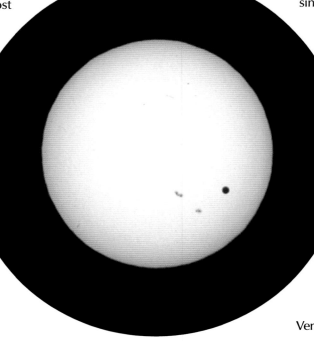

■ The occultation of a star by an orbiting planet brings about a small dip in the star's brightness. COROT will be looking for such "mini-eclipses." This artist's rendition shows an Earth-size planet transiting in front of a star.

to search for planets transiting stellar disks. When an otherwise invisible planet transits a star, a small drop in brightness should be observed: in effect a mini-eclipse. Events of this sort are not easily detectable with ground-based telescopes, but the concept was validated in 1999 when the transit of a giant exoplanet across the star HD 209458 was detected in this manner. Several additional giant planets have since been discovered this way. Although well over 130 exoplanets have already been discovered, these have all been giants detected indirectly by their gravitational tugs on the parent stars. Only massive Jupiter- and Saturn-size planets exert large enough effects on stars to be detected in this manner.

Since COROT should be two to three times more sensitive than conventional photometric telescopes, it should be capable of detecting not just giant planets but also Venus- and Earth-size planets. By monitoring some 50,000 stars over a period of five months, the mission will combine astroseismology and photometry, and hopefully detect planets as they transit their host stars. An Earth-size planet transiting a Sun-like star from an orbit similar to ours would cause a one ten-thousandth dip in the brightness of that star for a period of about five hours. Of course, in order to observe such an event, the planet, its parent star and the Earth would have to be in exact alignment, a statistical probability of about 1 percent. However, French researchers hope to significantly up the odds of success by surveying a very large number of stars to detect as many as 100 Earth-size exoplanets in addition to numerous larger ones.

If everything goes according to plan, the duration of the COROT mission might be extended, providing researchers with

■ *COROT will be able to continuously survey a field of 50,000 stars from its polar orbit. In addition to numerous gas giants, the spacecraft is expected to detect about 100 terrestrial-type exoplanets.*

five separate observing sessions. The initial survey should provide data on hundreds of transit events involving exoplanets of various types and sizes. The second survey will concentrate on planets with short orbital periods, and a third survey would be used to predict future transits of potential new planets. In each case, at least three successive transit events must be observed before it can be considered a genuine planetary crossing; others could simply be due to small variations in the star's normal brightness. The size, mass and orbital period (year) of an unseen planet can be derived by precisely measuring both the duration and magnitude of the drop in brightness of the star it transited.

The findings of the COROT mission will also play an important role in the development of additional planet-seeking space telescopes. If this little telescope does in fact discover several dozen or even hundreds of terrestrial-type extrasolar planets, that would clearly justify future space missions like Darwin and TPF, which are designed to directly image Earth-size exoplanets and undertake spectroscopic analysis of their atmospheres. ■

> An observing session

Time at the telescope is precious and sometimes hard to get. Astronomers are granted observing time at an observatory strictly on a merit basis and so must make the most of the opportunity and publish their results as quickly as possible. This is a competitive process repeated many times over and dictates how modern observatories operate.

TIME DEMANDS

Astronomers requesting telescope time must first prepare and submit an observing proposal. The project is then reviewed by a panel of experts who have the difficult task of selecting only the best proposals for telescope time. Requests are evaluated using the following criteria: scientific merit and relevance, timeliness and project feasibility. Typically, fewer than a third of all submissions are granted observing time. In case of rejection, the investigator or team must usually wait six months to a year before reapplying.

PREPARING FOR AN OBSERVING SESSION

Once telescope access time has been granted, the astronomer will usually spend several weeks preparing for the observing run. Typically this will include preparing an accurate list of objects to be observed and charts of the area of sky to be investigated, keeping in mind the number of nights or hours of telescope time allotted to the project. If space telescopes are involved or the use of one of the giant telescopes, then a minute-by-minute observing itinerary is required, outlining the most time-efficient sequence in which objects will be observed. Such "Phase II" observing logistics preparations are always done in concert with the agencies operating the telescopes in question.

A SESSION AT A MAJOR OPTICAL OBSERVATORY

Experienced astronomers visiting a major observatory, say in Chile, Australia or the United States, quickly become accustomed to the rules and routines at those sites, which vary little from visit to visit. They must get to the site quickly and contact the technical and support personnel at the tele-

scope to which they have been assigned, and then become familiar with all the instruments they will be using. The day before they start the observing run, the resident astronomer will outline how the telescope and support instrumentation are managed so that visitors can begin observing as soon as their session begins. The resident astronomer will usually be present during the first night of the run to ensure that the facilities function properly.

Visiting scientists usually arrive at their telescopes several hours before sunset to test and calibrate all equipment so they can start collecting data as soon as it is dark enough. Follow-up calibrations are also done the morning after an observing session.

● Obtaining the spectrum of a distant pulsar or quasar, or successfully imaging stars, nebulae and galaxies usually requires extensive preparation and planning on the part of astronomers. Reducing, analyzing and interpreting the data takes even longer and final publication of the results may involve years of work. The nebula HH-34 is shown above, taken with the FORS 2 imager-spectrograph at the focus of the 8.2 m Kueyen telescope.

● Observatories at remote locations, like Cerro Paranal, above, in the Atacama Desert, have on-site accommodations for astronomers, engineers and technical support staff. Facilities like this serve as science centers where researchers from around the world gather to discuss new observational techniques and theoretical aspects of their work.

Time passes quickly during an observing session, as the telescope moves from object to object collecting data. The entire operation is usually run by a night assistant thoroughly familiar with all instrumental idiosyncrasies. Visiting scientists direct the overall action of the telescope, except at really large facilities like the VLT, where a resident astronomer is always in charge. The latter works with the visitor and follows his directions, but also makes sure that the telescope and its instruments are not subjected to any potential operational risk. Many telescopes today are operated remotely, either from a nearby control center or from many miles away, thereby minimizing interruptions and the risk of stray light affecting instrument performance. Weather and temperature conditions, wind speeds and other information that affects the telescope are automatically fed to the control center, where computers compensate and/or adjust accordingly.

The morning after an observing run, the visiting astronomer must provide a report of the night's activities, including time spent observing, site statistics, any technical problems encountered, instrumentation settings or any other information that might be needed for the following night. A "day crew" will subsequently take over and repair or exchange instruments if needed.

At this stage, astronomers and night assistants usually take a well-earned rest. After waking sometime after noon, the visitor will generally take a first glance at the results of the night before, hoping that some exciting new discovery will pop right up. In the unlikely event that that is the case and that something new or unusual has been observed (a new comet, for example, or a supernova in another galaxy), other observatories can be alerted immediately by telegram or other means to follow up and confirm the discovery.

Later that afternoon, preparations are made for the second night of observing and any changes in scheduling,

instrumentation, calibration and so forth are discussed and implemented. This routine is followed for the duration of the visit. When the observing run is over, all data are copied and archived at the observatory for future reference and for researchers to work on at their home base.

FOLLOW-UP TO AN OBSERVING RUN

After returning to their home institutions researchers begin the cleanup and reduction of their "raw" data. This typically involves removal of any instrument-induced errors or observational artifacts to ensure the results apply only to the objects under investigation. This can be a rigorous and time-consuming process, often taking several months to complete, and it must be done in a timely fashion. Researchers have exclusive access to their results for one year, after which they must share them openly with other scientists. Once data reduction is completed, astronomers face the exciting task of interpreting them in theoretical terms or using them to propose new operational models. This is usually the most rewarding stage of the process.

PUBLISHING FINDINGS

After they have completed all data reduction and analysis, along with any novel interpretations or models, astronomers face the final stage of the process: writing a paper for submission to a peer-reviewed journal. This typically includes a concise yet comprehensive summary of the questions addressed followed by a description of the observations and the equipment and methods used to obtain them, the analytical methods used and the results obtained. This is followed by a discussion of the results and their significance in the broader context of the investigation. The paper is then submitted to an international journal for review by other experts in the field, whose job it is to evaluate the work objectively and, if warranted, recommend it for publication. ■

The Herschel Space Telescope

An Ariane 5 rocket will launch both the Herschel and the Planck space telescopes at Kourou in February, 2007. They will be placed into solar orbits at the Lagrangian point L_2, about 1.5 million km from Earth, to avoid intense infrared radiation from the Sun. It will take Herschel about four months to reach that point before beginning its three-year mission. With an aperture of 3.5 m, Herschel (originally named the Far InfraRed and Submillimeter Space Telescope or FIRST) will be the largest orbiting telescope until the deployment of the NGST. Its primary goals will be to study the formation and evolution of stars and galaxies.

Space-age technologies have finally put astronomers in a position to ask some very basic questions, questions they could not really answer prior to this, about the origins of stars and galaxies, two of the most important entities in the Universe. For instance, exactly when and how were galaxies formed? Much better mapping and distribution data is needed to answer this question, particularly for the most redshifted galaxies, those radiating strongest at longer wavelengths in the infrared and submillimeter range.

AT THE CORE OF STELLAR NURSERIES

Under what conditions and how are stars born? In very general terms we now understand this happens within the dense, cold and dust-rich globules of active nebulae, regions that are largely opaque to visible light. Infrared, submillimeter and radio wavelengths are the only way to probe deeper into these objects.

Unfortunately targets like these as well as distant galaxies are

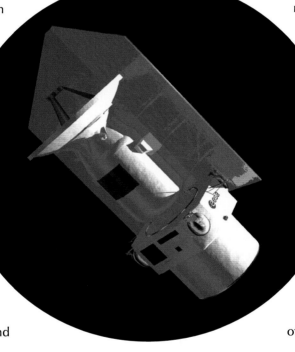

■ *The ESA's 3.5 m Herschel infrared space telescope will be 9 m long, 4.5 m in diameter and weigh 3.3 metric tons.*

dim and require large telescopes to collect enough light at the best wavelengths to reveal internal detail. Infrared and submillimeter astronomy is challenging with ground-based facilities. Even at Mauna Kea, with an elevation of 4,000 m, atmospheric heat and water vapor interfere with observations at these wavelengths. The infrared universe was really not fully revealed until satellites like IRAS and ISO were launched in the 1980s and 1990s and, in 2003, the Spitzer Space Telescope.

Named in honor of William Herschel, the British Royal Astronomer who discovered infrared light, the Herschel telescope will be larger and more versatile and placed in an orbit as far removed as possible from any interfering radiation by the Earth, Moon and Sun. The spacecraft will be under the direction of the European Space Agency, in partnership with NASA and several other research organizations.

With a 3.5 m diameter mirror, Herschel will be the largest submillimeter and far-infrared space telescope ever built, six times larger than IRAS and ISO and four times larger than the 0.85 m Spitzer Space Telescope. It will have greater light-gathering and resolving power than its predecessors. It will also cover the widest range at this wavelength, from 60 to 670 μm, and will be equipped with a battery of imagers and spectrometers.

The IRAS satellite could only observe at four bandwidths (12, 25, 60 and 100 μm) and the Spitzer telescope covers the 3 to 180 μm range. In short, the Herschel telescope should yield better results than any instrument to date, both in terms of resolution, spectral coverage and sensitivity.

■ *The ISO infrared satellite took this beautiful image of the Eagle Nebula, M16, at wavelengths of 7.7 and 14.5 μm. Image resolution is about 4 arcsec. The Herschel Space Telescope will attain comparable resolution, between 60 and 670 μm.*

In order for the Herschel Space Telescope to function to maximum potential, it must be equipped with the most advanced instrumentation and work under extremely cold conditions. In part this is achieved by placing it in an orbit as far away as is practical, at the Lagrangian antisolar point, 1.5 million km from Earth, where the ambient temperature is around –200°C. In addition, all instrumentation will be cooled inside a cryogenic unit to –271°C, less than three degrees above absolute zero.

Two instruments, PACS (Photodetector Array Camera and Spectrometer) and SPIRE (Spectral and Photometric Imaging Receiver) will carry out imaging and spectrometry, with average resolution between 300 and 3,000 km/s. PACS is optimized to work at the "shorter" wavelengths between 60 and 210 μm and SPIRE between 200 and 670 μm. PACS will provide higher-resolution images but SPIRE covers a wider field of view.

In addition these instruments are equipped with bolometric sensors for

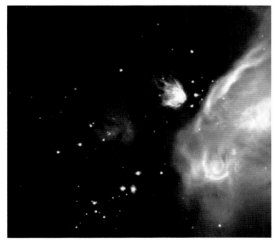

■ *Herschel will survey cold dark nebulae like this star-forming region around Rho Ophiuchus. The ISO telescope took this infrared image at wavelengths of 7.7 and 14.5 μm.*

photometry and spectroscopy, four with PACS and five with SPIRE. These sensors represent unique technologies on their own and will have to be cooled to 300 millikelvins to eliminate thermal interference from extraneous sources.

Thanks to receivers like the Heterodyne Instrument for the Far Infrared (HIFI), Herschel will be able to undertake ultrahigh-resolution spectroscopy (between 0.3 and 300 km/s) at two frequency levels, one between 157 and 212 μm, and the other between 240 and 625 μm. Unlike the other detectors aboard the spacecraft, HIFI is not an imaging instrument, but a combination of SIS sensors and hot electron bolometers.

The Herschel Space Telescope and future facilities like the NGST and ALMA are expected in combination to cover the full spectral range, from far-infrared to near-infrared and submillimeter radiation, and provide much needed information about the coldest and darkest reaches of the cosmos. ■

The Max Planck Surveyor

EUROPE

The Ariane 5 rocket carrying the Herschel Space Telescope into orbit will also launch a second satellite, the ESA's Max Planck Surveyor (Planck). Both observatories will be placed into solar orbits at the Lagrangian point L_2, about 1.5 million km from Earth on the side opposite the Sun. The Planck telescope must also end up as far away as possible from any Earth-based interference to carry out its mission, which is to measure minute fluctuations in the cosmic background radiation. This information is crucial for testing current theories on the formation and evolution of the Universe. Quietly following the Earth in its orbit around the Sun, Planck will record and monitor the oldest and most distant cosmic signals known.

In 1965, Arno Penzias and Robert Wilson, working with one of the first radio telescopes, discovered what was subsequently called the cosmic microwave background (CMB). This primordial radiation, manifest at very short wavelengths, represents the earliest "light" in the Universe when it was only 300,000 years old. It is in effect the "echo" of the big bang. As a first approximation, Penzias and Wilson estimated the temperature of this fossil radiation to be around 3 K, or about 3 degrees above absolute zero. They also saw it as isotropic radiation, meaning it came from every direction in the sky. This was an important finding since it validated the big bang theory. However, these observations did not explain how things like the first stars and galaxies formed, since that required at least local anisotropic variations in the CMB pattern. In other words, to account for large-scale structures, such as clusters of galaxies, a more heterogeneous temperature profile was expected. In 1989 the Cosmic Background Explorer (COBE) obtained extremely accurate CMB measurements of 2.726 K, but also hinted at

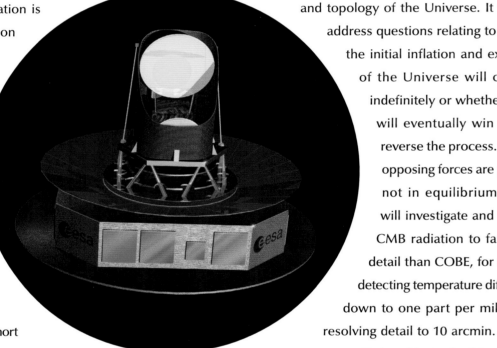

■ A single Ariane 5 rocket will launch both the Herschel Space Telescope and the Planck surveyor in 2007. Weighing close to 2 metric tons, Planck will be 3.8 m high and 4.5 m across.

minute variations or anisotropies in this fossil temperature pattern.

When deployed in 2007, the Planck surveyor will map CMB radiation in greater detail and at much higher sensitivity than ever before. This information has direct bearing on such fundamental aspects as the Hubble constant (the rate of expansion of the Universe), the cosmic distribution of dark matter and dark energy, the cosmological constant and theoretical considerations on the dynamics and topology of the Universe. It will also address questions relating to whether the initial inflation and expansion of the Universe will continue indefinitely or whether gravity will eventually win out and reverse the process. The two opposing forces are probably not in equilibrium. Planck will investigate and map the CMB radiation to far greater detail than COBE, for example, detecting temperature differences down to one part per million and resolving detail to 10 arcmin.

To be able to do this, however, Planck will have to be isolated from every conceivable source of microwave heat radiation, both internal and external, including not only the Earth, Sun and Moon, but also all instrument noise and background radiation from the Milky Way and other galaxies. In part this will be done through shielding and extensive cryogenic cooling to very low temperatures. Since it will not be possible to completely eliminate extraneous radiation from astronomical sources, Planck will map them accurately and subtract those readings from the microwave background signals. This will be done by taking multiple-channel readings to determine the level of signal contributed by the Milky Way and other galaxies against the background CMB radiation.

Great care was taken in designing the Planck space telescope

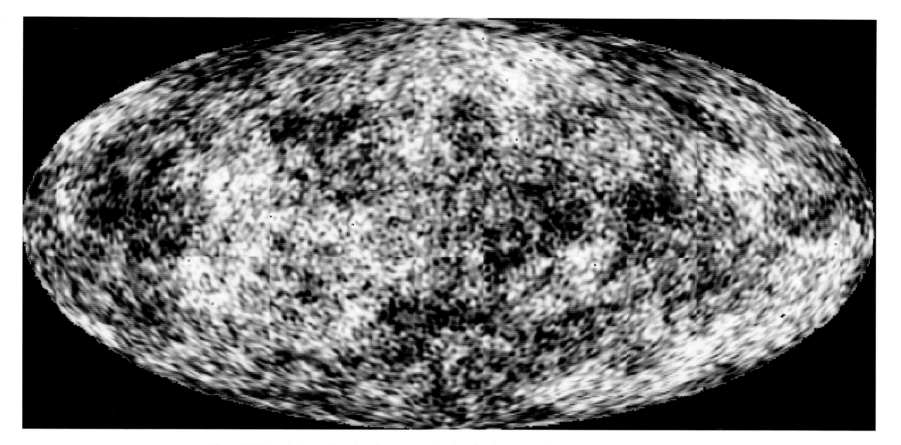

■ *After COBE and MAP, the Planck surveyor is the third space telescope exclusively dedicated to measuring fluctuations in cosmic background radiation. This computer-generated simulation illustrates how Planck is likely to portray the Universe just 300,000 years after the big bang.*

in order to ensure optimal performance level and success in this important mission. Its 1.5 m primary mirror will be able to record very weak signals. All its mirrors are made of lightweight carbon fiber with very high reflectivity. In addition, with angular resolution around 10 arcmin, Planck will reveal considerably more detail than COBE, which had a resolution of only 10 degrees and MAP, which had about a third of a degree.

COBE, MAP
AND PLANCK

Receivers working in the 25 to 900 GHz frequency range, which corresponds to wavelengths of 0.3 to 12 mm in the millimeter and submillimeter domain, typically include nine separate channels. Planck uses two types of receivers, one for low frequencies (30 to 100 GHz) and one for high frequencies (100 to 900 GHz). The low-frequency receiver uses 56 sensors equipped with high-performance transistors cooled to 20 K. The high-frequency receiver uses 48 bolometers cooled to less than one degree above

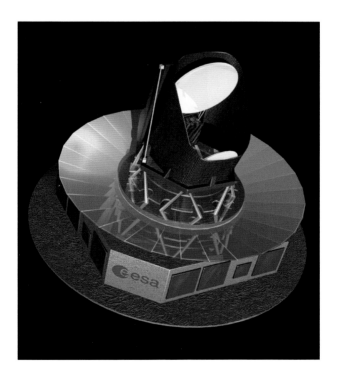

■ *Trailing the Earth from 1.5 million km, the Planck surveyor will be equipped with a 1.5 m mirror and scan the celestial sphere in the 0.3 to 12 mm wavelength range at better than 5 arcmin resolution.*

absolute zero. The first receiver will be built by the Institute of Technology and Research on Extraterrestrial Radiation in Bologna, Italy, and the second by a group led by the Institute of Space Astrophysics in Orsay, France. Apart from being a major scientific undertaking, the Planck project is a true adventure, involving about 100 scientists, engineers and technical staff, working at 40 different European and five American institutions.

Unlike space telescopes like Hubble, Planck will not focus on specific targets but survey the entire sky. With its optical axis offset 70 degrees relative to its axis of rotation, Planck will sweep the sky once a minute. By progressively adjusting its axis of rotation, the telescope will gradually cover the entire sky, and at the end of the 18-month mission Planck will have completed two full surveys. In addition to providing detailed information on small-scale anisotropies in the CMB radiation, Planck will provide data on the initial conditions in the cosmos that led to the formation of large-scale structures. ■

The James Webb Space Telescope

After 20 years of outstanding service at the end of this decade, the venerable Hubble Space Telescope (HST) is scheduled for decommission. Original plans called for its return to Earth via space shuttle, but NASA cancelled that in wake of the *Columbia* disaster. Sadly, Hubble is most likely now to meet a fiery end in the atmosphere. Plans for a successor to HST have been in the works for several years, resulting in a joint effort between NASA and the ESA to build a Next Generation Space Telescope (NGST). In 2002 the project was renamed the James Webb Space Telescope (JWST).

With the many advances in technology since the 1970s, when Hubble was designed, it is now feasible to build a far more advanced and powerful telescope. Currently on the drawing board is a 6.5 m infrared telescope to be launched in 2011, shortly after the HST is decommissioned. There are several historical reasons for choosing an infrared telescope. At the time Hubble was designed, infrared detectors were simply not available. The infrared end of the spectrum is of great interest to astronomers investigating the most distant objects in the Universe and tracking extrasolar planets. The JWST is projected to cost around US $1.2 million, considerably less than the estimated US $7 million for the HST program.

Launched by a standard Titan, Atlas or Ariane 5 rocket, the telescope will be able to operate around the clock from its 1.5 million km orbit at the L_2 Lagrangian point, where it will not be affected by radiation from the Earth, Moon or Sun. Facing away from these three bodies, it will have access to the entire celestial

UNITED STATES – EUROPE – CANADA

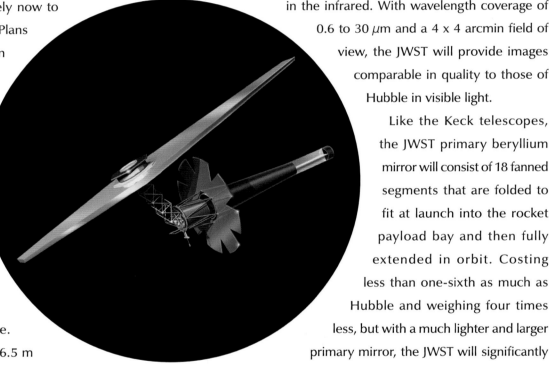

■ *The James Webb Space Telescope will be placed in a solar orbit 1.5 million km from Earth, putting it at a far better location than the 600 km high orbit of the Hubble Space Telescope.*

sphere over the course of a year. The L_2 point provides several additional advantages: the sky is much darker from that vantage point than it would be from Earth orbit, it is better shielded from the heat and light of the Sun, and the ambient temperature is extremely low, ideal for observing in the infrared. With wavelength coverage of 0.6 to 30 μm and a 4 x 4 arcmin field of view, the JWST will provide images comparable in quality to those of Hubble in visible light.

Like the Keck telescopes, the JWST primary beryllium mirror will consist of 18 fanned segments that are folded to fit at launch into the rocket payload bay and then fully extended in orbit. Costing less than one-sixth as much as Hubble and weighing four times less, but with a much lighter and larger primary mirror, the JWST will significantly outperform it as well.

Over the past decade Hubble has surveyed galaxies with redshifts between 2 and 6, or as far back as their birth 12 billion to 14 billion years ago, and provided us with a general view of their evolution and distribution. With far greater sensitivity and penetration power, the JWST will extend its survey to objects with redshifts between 6 and 10, when the Universe was less than a billion years old. This should provide much needed information about the origin of the earliest known objects in the Universe, the primordial clouds of gas and stars that eventually coalesced into galaxies.

Like the Hubble Deep Field Survey that observed 30th magnitude objects, the JWST will undertake long exposure probes of selected regions of the sky. Shooting through eight different filters

■ *This simulation of a JWST deep-field image, taken through eight infrared filters (from 0.6 to 10 μm), illustrates what might be expected after an exposure of 1,350 hours. Objects down to 33rd magnitude and redshifts of at least 10 will be recorded.*

to cover the full 0.6 to 10 micron range, this will require a total exposure time of 1,350 hours — nearly six months of continuous observations — in order to reach 33rd magnitude objects.

A second major objective of the new space telescope will be a systematic survey of distant supernovae with redshifts between 1 and 5, to determine whether their light curves are similar to those of supernovae closer to us. This information will help more accurately define the Hubble and cosmological constants and the critical density of the Universe. More accurate determination of these three parameters will help cosmologists decide which models of the cosmos predicted by general relativity most accurately describe the real Universe. It seems likely that in the coming decade we will fully understand what conditions were like at the time immediately after the

big bang, and the true shape and ultimate fate of the Universe.

OBSERVING UNDER IDEAL CONDITIONS

In addition to addressing questions about cosmology, the James Webb telescope will actively search for exoplanets. This will include a systematic survey over several months of the closest 200 stars within a radius of 25 light-years from Earth. With magnitudes between 25 and 30, most exoplanets cannot be detected by ground-based telescopes, since the stars they orbit are hundreds of millions times brighter than they are. This ratio is reduced to less than one million at 5 μm in the infrared. From its near-ideal vantage at L_2, JWST cameras with stellar coronagraphs should be able to detect all giant gas planets around these stars, planets in the Jupiter- to Saturn-size class. ■

> Data reduction and analysis

*T*o derive reliable and error-free information, raw data collected at a telescope must be reduced and normalized by removing noise and artifacts caused by poor seeing conditions or by the instruments themselves. Only in this way will the results obtained be dependable and consistent.

AN OVERVIEW OF THE PROCESS

Charge-coupled devices (CCDs) are special computer chips that collect photons and convert the energy into electric signals. These signals are subsequently amplified and converted into a series of numbers stored by computers. Each CCD exposure or image is stored as an array of numbers corresponding to data from every pixel in the chip. In addition to astronomical information, however, the data usually also contain errors and artifacts introduced by atmospheric distortion, optical effects, instrument noise, electrical problems and various other factors.

Because of this, no matter how good a raw CCD image might appear, it has to be "cleaned-up" and processed. This is really the most important step in data reduction. Until all background noise has been removed, and both the number of photons per pixel and the amount of sky covered by each pixel have been reliably established, the image is of little scientific value. Such image processing is typically done through a combination of photometric and dimensional calibration procedures and spectral adjustments.

Most data normalization procedures are pretty standard now, which helps produce reliable and comparable results. It's not possible to totally eliminate noise and artifacts from the results, particularly things like "photon noise," which is produced by the target objects themselves, or spontaneous instrument noise.

READOUT NOISE AND DARK CURRENT

When CCD images are downloaded an effect called "readout noise" is added to the data. This effect can be averaged for

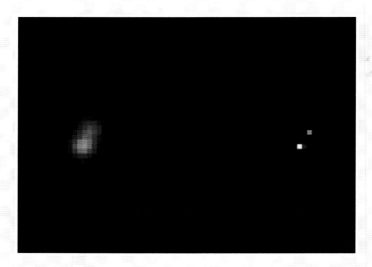

● The binary star system GJ 263, imaged by the 8.2 m Kuyen telescope with the NAOS adaptive optics system, is shown on the left. Separated by just 0.03 arcsec, the pair is just barely split. The same two stars are shown clearly split on the right, following extensive image processing.

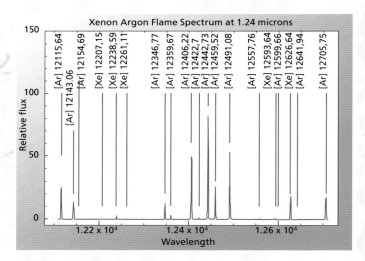

● This graph shows a portion of the infrared spectrum between 1.21 and 1.27 microns, a mixture of argon and xenon. Reference spectra like this are used to identify elements in emission spectra of astronomical objects.

● This photograph of M104 in Virgo was taken by the FORS 1 spectrograph-imager of the 8.2 m Antu telescope. The image was compiled from three exposures at 554, 657 and 768 nm, for a total time of eight minutes. Image resolution is close to 0.6 arcsec. This giant spiral galaxy is about 50 million light-years distant.

a given instrument (offset noise) and suppressed by taking a "dark frame" exposure, which is then subtracted from the informational signal to provide a corrected final image.

A second factor that must be taken into consideration with CCDs is "dark current," caused by random migrations of signal between pixels. This type of background noise builds up over time and can affect imaging of faint objects at the limit of the receiver's sensitivity. To overcome this problem, dark frames are taken with integration times similar to the original exposure, and then subtracted from the data.

FLAT FIELDING

Pixels in any CCD array do not respond homogeneously across the entire field and can result in irregularities that must be corrected. One way to do this is to illuminate the detector with light reflected off a white screen. The resulting image, called a flat field, highlights areas on the chip with unevenly responding pixels and other irregularities. They can be corrected by dividing the pixel values of the original image corrected for readout noise by those of the flat field.

ELIMINATING BAD PIXELS

A final step in CCD data reduction is to eliminate pixels that are "dead" or do not respond in linear fashion. Their signals must be eliminated because they contain no information or worse, they add erroneous information. Dark frames and flat fielding readily identify abnormal pixels that can then be tagged for future reference and removed by special software algorithms.

Most CCDs also generate artifacts called "cosmic rays." These appear as random points of light in isolated pixels and increase in number with longer exposure times. Since these are unpredictable events, they cannot be treated like dead pixels and routinely eliminated from the data. Instead every frame must be processed to show abnormally responding pixels, and their signals averaged with respect to adjacent pixels.

PHOTOMETRIC CALIBRATIONS

At this stage images are essentially free of most residual instrument noise and other artifacts, and contain only "true" information about the object observed. However, these data are still encoded as arbitrary Analog to Digital Units (ADUs) and must be converted to scientifically meaningful units like photon flux. The number of ADUs in a given frame will depend not only on the specific object imaged, but also on how high up it was in the sky, the amount of atmospheric absorption during the exposure, the type of detector used, the efficiency of optical transmission, and so forth. To compensate for these variables and standardize their results, astronomers always calibrate their receivers by measuring the brightness of adjacent reference stars whose energy output is very stable and well known. Without this type of photometric calibra-tion, ADUs cannot be reliably converted into quantitative units such as photon flux.

ANGULAR CALIBRATIONS

Lastly, the actual field of coverage (or arc seconds per pixel) in a CCD chip must be accurately determined. In theory this is easily calculated once all optical elements and the total area of the CCD are factored in. In practice, however, this is not accurate enough for quantitative purposes nor is the approxi-mate field orientation indicated by the telescope's guidance systems. Both parameters can be accurately established using well-characterized binary stars or similar objects as standards. When these final calibrations are done, image data reduction is completed. Spectroscopic data must be calibrated and stan-dardized in a different fashion.

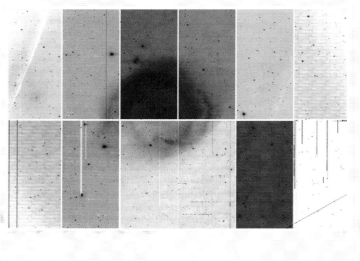

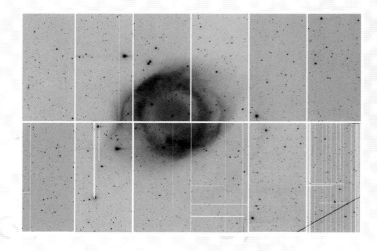

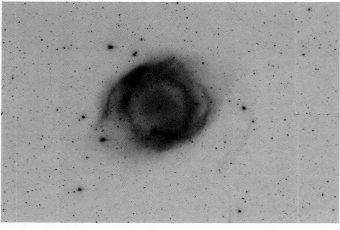

● The above illustration shows four stages in processing an image of the Helix Nebula, taken with the CFH12K camera and the 3.6 m CFHT. The raw image at top left is shown as it appeared on the 12 individual chips of the CCD mosaic. Each processing step provides useful scientific information about this object, imaged through different wavelength filters and then combined to produce a final true color image.

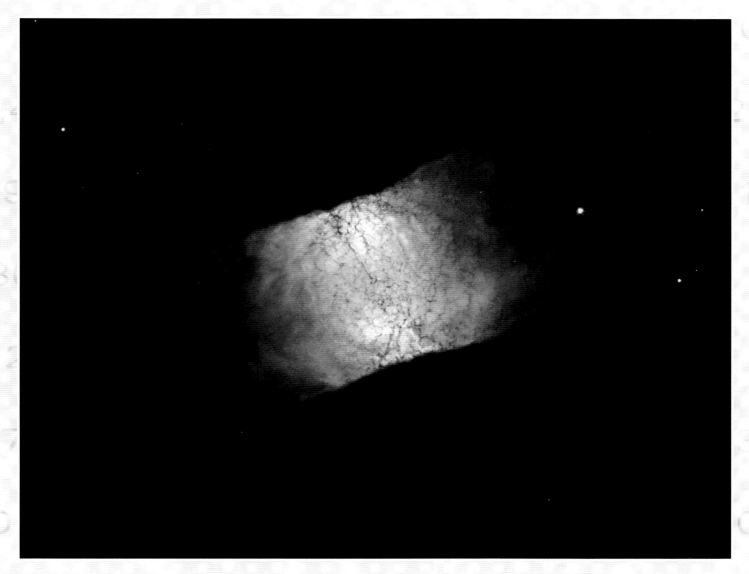

● This image of the planetary nebula IC 4406 was taken with Hubble's WFPC2 imager. This nebula, located about 1,900 light-years away in the constellation Lupus, is about one light-year across, and was formed several thousand years ago by an aging red giant that blew away its outer layers. This one-hour exposure was taken through 502, 656 and 658 nm filters.

SPECTROSCOPIC STANDARDIZATIONS

The whole purpose of spectroscopy is to disperse light and provide information in basically two axes: a vertical or spatial axis and a dispersion or wavelength axis. For any given detector, wavelength calibration is carried out using the emission lines of known standards like xenon and argon to determine the number of angstroms or microns per pixel. Once that is done, this value can be extended across the full array of pixels in the instrument and provide an internal standard of reference for astronomical spectra.

IMAGE PROCESSING AFTER DATA NORMALIZATION

Once the results have been normalized as much as possible, as indicated above, they can at last be evaluated quantitatively. This may require additional processing to sharpen or bring out hidden image details through such procedures as "deconvolution" and deblurring. This does not introduce information that is not there, but enhances information inherent in an image that may have been distorted by the telescope or receiver. Deconvolution and similar image-sharpening techniques are based on complex mathematical algorithms.

THE TOOLS OF DATA PROCESSING

All the data reduction processes described above are computer based and use algorithms specifically developed for astronomical purposes. They include a wide range of tools to generate images with quantitative information about the objects observed. Most of the larger observatories today are fully set up for data reduction and processing, although astronomers often develop their own specialized analytical methods themselves. ■

The Space Interferometry Mission

U N I T E D
S T A T E S

I n the late 1990s the National Aeronautics and Space Administration approved funding for the Space Interferometry Mission (SIM). Slated for launch in 2009, this instrument (built by the Jet Propulsion Laboratory) will be NASA's first large space interferometer. Based on the classic design pioneered in 1919 by Albert Michelson at Mount Wilson Observatory, SIM will consist of a 10 m long platform supporting three pairs of .3 m mirrors. After transport into a low, 400 km Earth orbit by the space shuttle, the spacecraft will use its own powerful rocket engine to propel it into a much higher orbit. Following the Earth around the Sun, it will gradually move outward at about 15 million km per year, reaching a distance of 100 million km toward the end of its mission in 2015. Facing away from the Sun and supported by a large array of solar panels, SIM will be in continuous operational mode.

■ *The Space Interferometry Mission (SIM), scheduled for launch in 2009 aboard the space shuttle, will gradually move toward a final orbit some 100 million km from Earth.*

Optical Interferometry, page 104). With this technique the spacecraft can investigate circumstellar environments with unprecedented clarity. By attenuating light intensity ten thousandfold, or nearly 10 magnitudes, researchers hope to detect the zodiacal glow around sun-like stars. Such diffuse gas rings are characteristic of planetary systems like ours, and if they are dense enough they could impede future efforts to directly image planets around them.

With its .3 m aperture telescopes, SIM will be able to record 8th magnitude stars in about a minute, but will need more than eight hours exposure to record 20th magnitude objects, its theoretical limit. Imaging, however, is not the spacecraft's primary mission. Instead, it will be the most accurate astrometric instrument ever built. Using the database provided by the Hipparcus spacecraft in the 1990s, which cataloged stellar positions to within 0.001 arcsec, SIM will be used to measure stellar positions and distances with unprecedented accuracy.

THE ZODIACAL LIGHT

As a classic optical interferometer, SIM will operate in the 500 to 900 nm wavelength range. It will require many hours of exposure time and cover only a tiny 1 arcsec area of sky, but at very high 0.01 arcsec angular resolution. One pair of mirrors will keep the interferometer aligned and one pair will be used for calibration. The other telescopes will gather data from various angles along its extendable 1 to 10 m baseline.

SIM will also be the first space telescope to use optical nulling to attenuate starlight through destructive interference (see

The interferometer will continuously adjust its reference points and measure parallax angles to stars and quasars with an accuracy of 0.000,001 arcsec (or 1 micro arcsec). Parallax is used to calculate both the distance and proper motion of celestial objects, and SIM will be able to do this with any stars within a radius of more than 500,000 light-years, which encompasses the Milky Way and its satellite galaxies.

Special emphasis will be placed on accurate distance measurements to such standard candles as Cepheid and RR Lyrae variables,

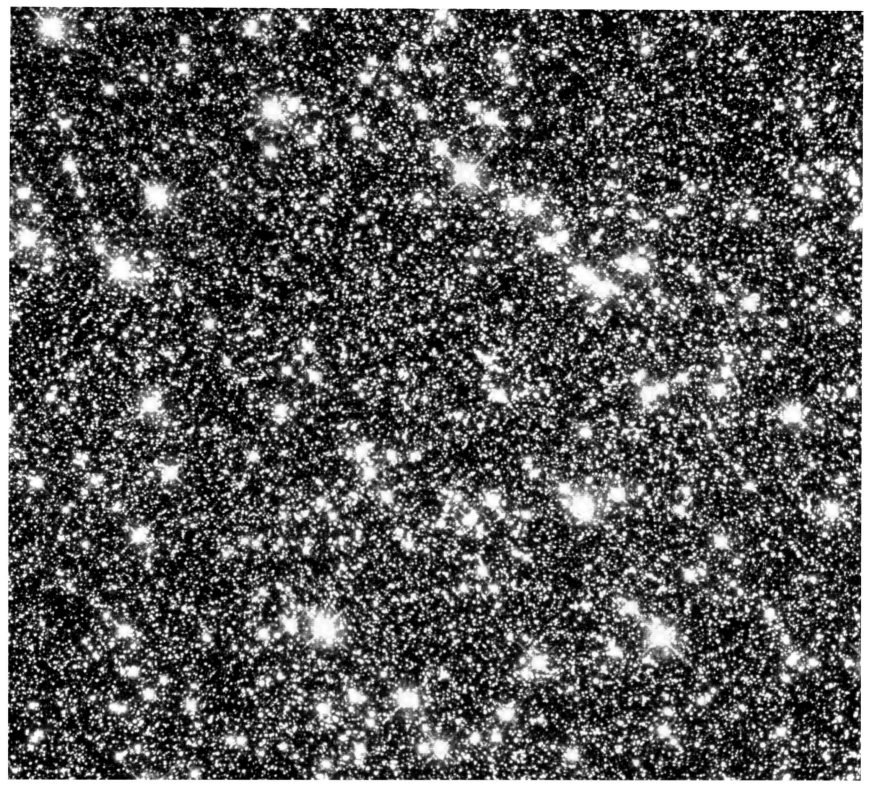

■ *SIM will be able to measure distances, proper motions and absolute magnitudes of any star in or near the Milky Way and its satellite galaxies.*

red giants used as distance and recession indicators to galaxies as far out as 100 million light-years. SIM will not only obtain highly accurate information about physical characteristics, diameter, mass and luminosity of such stars, but also data on the overall mass and distances of all globular clusters orbiting the Milky Way.

Closer at hand, the spacecraft will also be searching for extra-solar planets. Like other planned space telescopes, including COROT, Kepler and Eddington, SIM will be looking for Earth-size planets. It will do this indirectly: not by imaging, but by looking for gravitational wobbles in the motion of nearby stars. With Sun-like stars located less than 30 light-years from us as the prime targets, SIM will be sensitive enough to detect between 1 and 10 Earth masses.

The success of this effort will influence to some degree future plans to find habitable planets in the Universe. It will serve as a prototype for even larger spacecraft, potentially capable not only of identifying other "blue" planets like Earth, but of analyzing their atmospheres and actually imaging their surfaces. ■

The Gaia Space Observatory

As the most ambitious celestial cartography project ever planned, the European Space Agency's Gaia observatory probably represents the pivotal space mission of the 21st century. Its basic goal will be nothing less than measuring the positions, distances, luminosities and proper motions of about one billion stars in our galaxy.

Almost every known physical and quantitative stellar parameter will be recorded with extreme precision by this extraordinary spacecraft. By measuring the precise apparent brightness and distance of stars through 11 different optical filters, Gaia will also directly determine the absolute brightness, surface temperature, size, mass, internal workings, energy output and even the age of every star it observes. Such fundamental astrometric work was first undertaken in 1990 by the European Hipparcus spacecraft, which surveyed the position and proper motions of some 118,000 stars in our galaxy to within 0.001 arcsecs out to about 3,000 light-years.

Gaia will completely revolutionize such efforts by mapping the galaxy in three dimensions for the first time, and obtaining accurate measurements of its total mass, the differential motion of various regions within it and the physical characteristics of huge numbers of stars. This megasurvey will include nearly 1 percent of the galaxy's total stellar population. Every aspect of modern astrophysics will be affected by this important mission, including such fundamental features as the curvature of space and similar relativistic parameters.

Slated for launch by an Ariane 5 rocket in about 2010, the 3-metric-ton spacecraft will be placed into a stable solar orbit at the Lagrangian L_2 point some 1.5 million km from Earth. From there Gaia's two large telescopes will systematically scan the celestial sphere from opposing angles 106° apart. By tracing large

EUROPE

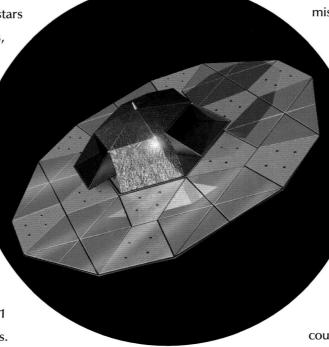

■ *The European Gaia spacecraft is scheduled for launch around 2010 into a solar orbit 1.5 million km from Earth at the L_2 Lagrangian point. Its survey of the Milky Way will take at least five years.*

circles across the sky with its two optical sensors, the spacecraft will act somewhat like a compass and scan each star several hundred times from different directions. Very accurate positional and luminosity data will be obtained this way as stars slowly drift across each telescope's CCD sensor. Real-time data reduction will be done aboard the satellite and the results relayed to ESA computers for storage and further analysis. The end result will be a whole sky map based on several billion separate star measurements.

The operational underpinnings of this mission are parallax angle measurements. When you observe a nearby object, a church steeple for example, against a distant group of mountains, and then shift position slightly to the left or right, the steeple also seems to change position against the background. This effect, known as "parallax," also works for stars, only on a much, much smaller scale, since to the naked eye stars all appear fixed at the same distance from us. In reality of course some stars are just a few light-years away and others thousands and millions of times further.

Stellar parallax becomes apparent when we observe a relatively close star six months apart, when the Earth is at opposite positions in its orbit. A minuscule shift against more distant background stars can be measured. The star's distance can then be calculated by simple triangulation, using a 300 million km baseline and the stars' apparent displacement angle.

Star parallax angles are always very small and measured in arc seconds. All parallaxes measured by Gaia will fall between 0.742 arcsecs for stars closest to the Sun, and 0.000,01 arcsecs (10 micro arcsecs) for those furthest away. The latter is equivalent to the width of a hair seen from 1,000 km away or a marble at the distance to the Moon.

■ *Gaia will be the successor of the Hipparcus spacecraft, which obtained positional measurements of some 118,000 stars in the 1990s to an accuracy of about 0.001 arcsecs. Gaia will observe one billion stars with a hundred times greater accuracy.*

OBSERVING A BILLION STARS

One limitation facing Gaia is that closer stars with the largest parallaxes can be measured with greater accuracy than stars further away. Since the spacecraft will not be able to measure angles smaller than 10 micro arcsecs, its distance limit will be about 100,000 parsecs (parsecs are simply a distance given by the inverse of the parallax angle), or about 326,000 light-years.

Gaia's twin telescopes will contain rectangular mirrors measuring 1.7 x 0.7 m. These will focus light onto huge CCD cameras composed of 136 chips with 2,780 x 2,150 pixels each, for a total mosaic of 800 million pixels! Gaia's extraordinary astrometric precision will be attained through numerous repeat measurements of stellar positions during the spacecraft's five-year operational lifetime. It will be possible, for instance, to determine distances to such nearby stars as Proxima Centauri, Sirius, Procyon, Altair and Vega to within a billion kilometers.

This information will be invaluable for astronomers. Every star to 20th magnitude will be surveyed, including 20 million distance measurements out to 10,000 light-years, with better than 1 percent accuracy. Still further, distance measurements accurate to better than 10 percent will extend to about 100,000 light-years. This is possible because stars surveyed by Gaia will exhibit some proper motion over the mission's five-year duration no matter where they are located within our galaxy.

This will permit us at last to fully delineate the size, mass and exact shape of our galaxy, and obtain invaluable information about the exact masses of the estimated tens of millions of double stars to better than 1 percent. In addition, absolute magnitudes will be determined of such standard distance indicators as red giant Cepheid variables and RR Lyrae stars. Equally important, Gaia will obtain similarly accurate measurements of standard candles in the nearby Magellanic Clouds, the Milky Way's two principal satellite galaxies.

One of Gaia's primary tasks will be to completely map the Sun's proximal stellar neighborhood out to about 500 light-years. No object within that space should escape detection, including all red, brown and white dwarfs, as the distance to every single star as well as every gas planet larger than Jupiter will be determined to within a few light-days! In fact the survey is expected to discover around 30,000 new giant exoplanets.

In addition to cataloging more than a billion stars, this mission is expected to discover a million new quasars, some hundred thousand extragalactic supernovae and close to a million new asteroids within our solar system. As well as laying the foundation for a comprehensive galactic encyclopedia, Gaia will provide data for future generations of researchers well before more ambitious astrometric missions are deployed to map the entire local cluster of galaxies. ■

The Laser Interferometer Space Antenna

As the largest scientific instrument contemplated to date, the Laser Interferometer Space Antenna (LISA) is a collaborative effort involving physicists and astronomers from NASA and the ESA. Like Virgo and LIGO, two similar ground-based facilities in Europe and the United States, instead of electromagnetic radiation LISA is designed to detect space/time distortions or the gravity waves predicted by Einstein's theory of general relativity. If funding is approved for various components of the project, NASA and ESA plan to launch the spacecraft into Earth orbit in about 2011 using a Delta IV rocket. LISA actually consists of three individual spacecraft, each of which will propel itself into a solar orbit trailing the Earth by about 50 million km. After several months, once they have arrived at their destination, the three spacecraft will separate into an equilateral triangle, about 5 million km per side. Each spacecraft will carry two .3 m diameter mirrors equipped with powerful 1-watt Neodyme-YAG lasers emitting light of 1 μ wavelength. Functioning both as laser emitters and receivers, each telescope will be precisely aligned with its distant neighbors into the largest-ever Michelson interferometer, whose three components will be in constant communication through overlapping laser beams.

Each spacecraft will also carry a reference or test mass of several kilograms that is directly linked to the lasers. A small change in distance between any two of the test masses will produce a phase shift between the outgoing and incoming beams and signal passage of a gravitational wave. It is estimated that LISA

■ *Astronomers have been studying the effects of space/time curvature as predicted by general relativity for more than 20 years. One such effect is illustrated here through gravitational lensing of objects lying behind the galaxy cluster Abell 2218.*

will be able to detect changes as small as 10 picometers (about one-tenth the size of an atom) every second!

Maintaining such extraordinary precision is not the main challenge that LISA designers face, however. Their biggest concern will be how to effectively isolate the three spacecraft and limit extraneous effects that could impact LISA's performance. Sunlight, for example, could cause slight shifts in the position of the three spacecraft. To minimize such effects, the reference test masses will be isolated deep within each spacecraft and electrodes will constantly monitor their electrical capacitance. Should there be a change in capacitance indicating a slight shift in position, corrections will be made through liquid cesium motors that can apply thrust levels as low as 0.1 to 100 micro-newtons.

LISA will be several orders of magnitude more sensitive than any of its ground-based counterparts because it has three very long branches, instead of the two 3 km long arms of Virgo, for example. This will permit LISA to accurately pinpoint the source of any gravitational waves it detects. In addition, as it orbits the Sun, gradual changes in the orientation of LISA's 5 million km triangular configuration will allow it to locate such sources with even greater precision.

HOW DO BLACK HOLES FORM?

Let's assume that a pair of neutron stars emits intense gravitational waves at regular intervals. Moving at the speed of light, it would take more than 16 seconds for these waves to traverse

the full LISA array. Over the course of several months, researchers would be able to trace their exact direction and point of origin by noting the precise moment each wave intersects all three laser beams.

LISA is designed primarily to investigate black holes and their surroundings. It will do this in part by monitoring the cores of as many galaxies and quasars as possible, many of which contain giant black holes with masses of 1 million to 100 million suns. Astronomers will look for any sudden gravitational perturbations resulting from stars crossing the event horizon of these black holes. LISA will also monitor other more classic gravity-wave emitters such as binary pulsars, gravitationally-linked pulsars and black holes, and rapidly rotating paired black holes. Many of these bodies release a portion of their energy as gravitational waves, and are destined to merge, resulting in cataclysmic explosions. Mergers of this type, which probably give rise to supermassive black holes, have been observed as gamma-ray bursts like those detected by the HETE spacecraft. However, reconstructing events where as much energy is released in less than a minute than the Sun has released in its 10 billion years of existence is extremely difficult, since they are invariably observed after the fact. LISA will be key to the study of such mergers, as it will complement observations by instruments working at other wavelengths. ■

> Astronomical charts and catalogs

*A*ncient astronomers quickly learned the value of cataloging and classifying celestial objects. This is perhaps even more important today, particularly as the volume of astronomical information and data increases exponentially. Accurate classification of celestial objects is of paramount importance, since the tools to access hundreds of catalogs and databases are now available to astronomers worldwide.

WHY INVENTORY CELESTIAL OBJECTS?

Astronomers have charted the positions and cataloged the brightness of celestial objects since antiquity. Our ancestors learned early on that to understand the mysteries of the sky above them, they had to first chart the positions and movements of the stars and the planets. This is still the case today, only on a much larger scale. Astronomers now strive not only to map every object in the sky, but also to clarify the processes and dynamics governing their development and evolution. These are complex and challenging tasks. Since the results often depend on difficult and costly observational procedures, they become that much more valuable to the astronomical community. Today researchers worldwide have access to and rely on observations often accumulated over many decades and even centuries. Who could have imagined, for example, that some of the observations made by ancient Chinese astronomers would still be of value today? It is clearly of utmost importance, therefore, that all observations and data be carefully recorded and cataloged for future generations.

THE DEVELOPMENT OF STAR CHARTS AND CATALOGS

The *Almagest* was one of the great astronomical texts of antiquity. Based entirely on naked eye observations, it listed all prominent constellations and about a thousand assorted celestial objects. It was also used for nearly 1,400 years!

The invention of the telescope in the 16th century changed all that, however, and by the end of the 19th century the number of new objects cataloged had risen to several hundred thousand stars and "nebulae." Today, as ever-larger telescopes become available, countless newly discovered galaxies, pulsars and quasars must also be mapped and cataloged. Modern databases now often include information on these objects as they appear at wavelengths other than the visible range. Cataloging on this scale will increase even more in the future, as large survey telescopes equipped with sensitive widefield detectors complete their work.

HOW ASTRONOMICAL DATA ARE ARCHIVED

With the dramatic accumulation of new astronomical data in the latter half of the 20th century, scientists had to develop novel methods of information storage, archiving and retrieval. Space observatories really led the way here by first recording observations on magnetic tape and then systematically archiving them on optical media. After they were fully processed the data were made available to the wider scientific community. Because of this, databases from spacecraft like COBE, HIU, ROSAT, Compton, the HST and others are valued sources of information for research, discovery and publication.

It took considerably longer for ground-based optical observatories to provide ready access to their extensive archives, which consisted for the most part of libraries of photographic plates that were far more difficult to catalog and process than the well-indexed data sent back by spacecraft. The major ground-based observatories today (ESO, Gemini, CFHT, etc.) use digital cameras and spectrometers, with facilities to directly archive all observations on site. Researchers world-

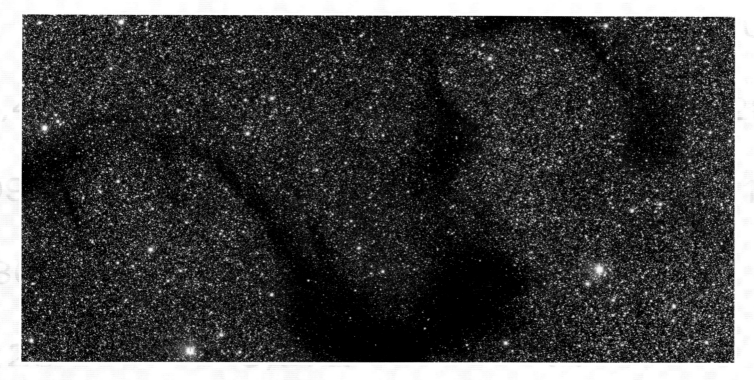

● The *Almagest*, a star catalog compiled in 200 B.C. by Greek astronomer Hipparcus, listed slightly more than a thousand stars. At the end of the 18th century Lalande's star atlas listed 47,390 objects, and by the late 20th century the Tycho atlas listed over a million. In the coming decade, the Gaia spacecraft is expected to catalog more than a billion stars.

wide can readily access these facilities, creating in effect a huge virtual observatory.

DATABASE ARCHIVES

As more and more observations and data accumulate and increasingly detailed sky surveys are undertaken, careful classification and analysis is more important than ever. That is why hundreds of different astronomical databases have been compiled, many listing only certain types of objects or information. For example, there are lists of all known quasars with their positions and magnitudes, and catalogs of all known double stars, with information on their position, angular separation and the magnitude of each component. These kinds of records are really extensions to the classic *Almagest*, but of course far more extensive and informative.

Other types of databases encompass all the objects recorded in systematic sky surveys, including those carried out by ground-based telescopes and spacecraft. Examples here are the 1983 IRAS Survey, listing the positions and infrared magnitudes of more than 200,000 objects, and the Hipparcus Survey, listing the position and proper motion of several hundred thousand stars observed between 1989 and 1993.

DATA COORDINATING CENTERS

For several years now dedicated data repositories have provided astronomers around the world with direct online access to vast banks of information and observations. Centers like these have become indispensable resources for research. One of these

centers, in Strasbourg, France. provides immediate access to literally hundreds of indexed databases. These can furnish information on the physical characteristics and quantitative aspects for any object in the database, as well as references to other relevant publications. Similar centers in Canada (CADC), the United States (ADS) and at NASA (ADC) are all closely linked to provide a planetwide information network for researchers. ■

● In the past 100 years the number of objects recorded by astronomers has increased exponentially. This small 7 x 7 arcmin field in Pisces shows hundreds of galaxies to 29th magnitude. The 70-hour exposure was taken with the 4.2 m Herschel telescope. The full celestial sphere contains more than 100 billion galaxies as faint as 30th magnitude.

Darwin and the Terrestrial Planet Finder

Following the Sun some 1.5 million km from Earth, the Darwin interferometer is looking for evidence of biogenetically produced oxygen in the atmosphere of an Earth-like planet orbiting another star. Is this science fiction? Not really. Just a few years ago the notion of looking for distant terrestrial planets that might be harboring life was indeed little more than a dream. Now thanks to enormous technological advances in interferometry, this may well happen before 2020. It will not be an easy undertaking. It will require an array of powerful space telescopes equipped with advanced technology to suppress the blinding glare of stars that obscures any dim planets circling them. This is precisely what the ESA's Darwin and NASA's Terrestrial Planet Finder (TPF) missions will attempt to do around 2015.

The discovery of the first extra-solar planets in 1995 was a milestone in the annals of astronomy, since it confirmed that our solar system is not unique and that giant planets do orbit other suns. These planets were not detected directly of course, but indirectly through gravitational tugs exerted on their parent stars. With this breakthrough astronomers were ready to exploit entirely new strategies in searching for planetary systems beyond our own. The first approach will use stellar coronagraphy and ground-based interferometry for direct visualization of gas giants like Jupiter and Saturn. The second approach will rely on space missions like COROT and Kepler, and use indirect methods to detect Earth-size, rocky planets. The third phase will attempt to image

INTERNATIONAL

■ *This Hubble image shows the red dwarf GL 623 B, which is separated by 0.25 arcsecs (300 million km) from the star GL 623 A. The former is 125 times dimmer than the latter. In comparison, the planets that the Darwin Space Interferometer will seek are a million times dimmer than the stars they orbit.*

terrestrial planets directly and look for signs of biological activity. That of course would be a pivotal discovery, since humans have speculated about the potential universality of life for centuries.

Terrestrial planets are much more difficult to detect because they will be a thousand times dimmer than giant gas planets and about a billion times fainter at visible wavelengths than the stars they orbit. Fortunately this contrast ratio is lowered to about a million to one at mid-infrared wavelengths. Another hurdle, however, is that any potential life-bearing terrestrial planet must be close enough to its parent star to retain water in liquid form. Moreover, in a system similar to ours located 10 parsecs away (about 32 light-years), an Earth-like planet located at 1 AU from its parent star would display an angular separation of only 0.1 arcsecs from our vantage point.

Any optical system hoping to image an Earth-like planet directly would need both very high resolving power and exceptional sensitivity over a wide dynamic range. It would also need high-resolution spectroscopic capabilities to detect what are presently considered the most likely chemical signatures for life: atmospheric oxygen and ozone. Ozone is generated through oxygen, which is itself the product of photosynthesis, as carried out by plants and micro-organisms. At present, we know of no other mechanism beside photosynthesis that could generate and release large amounts of oxygen into a planet's atmosphere. Both oxygen and ozone display characteristic spectral signatures near 10 μm in the infrared.

■ *The ESA's proposed Darwin project consists of an array of six 1.5 m telescopes forming a 50 m diameter circle around a central beam-combining telescope. A separate satellite relays communications with Earth.*

THE QUEST FOR OTHER EARTHS

Only a large array of space telescopes separated by several dozen meters will have the sensitivity and resolving power for such a search. As with all interferometry, the greater the number of individual telescopes involved, the greater the array's imaging capabilities, but also the higher the cost. It is estimated that with a reasonable budget, an array of four or five 2 to 3 m telescopes could both image Earth-size planets directly around nearby systems (within a few tens of light-years) and also obtain spectra in a reasonable period of time. In practice this means that planetary imaging would require exposures lasting several dozen hours and usable spectra with exposure of several weeks!

Although Darwin and TPF are both intended to meet these challenges, neither project has been finalized as to the number, size and configuration of telescopes to be deployed. Whichever designs are ultimately adopted, they will face some formidable technical challenges, including the deployment and control of a number of telescopes spaced dozens of meters apart, yet pinpointed collectively to within a fraction of a micrometer! An undertaking of this magnitude will obviously require several smaller trial missions first.

A second major challenge is how to effectively attenuate the light of stars with prospective terrestrial planets in order to image these much dimmer bodies. This will really hinge on the development of better "nulling" techniques to cancel out starlight, work presently underway with the Keck and VLT interferometers.

Like the James Webb Space Telescope, Darwin will observe in the infrared range. Since this will make it highly sensitive to thermal interference, the interferometer must be placed as far as possible from Earth and its atmosphere.

Since the two planet-finder projects have similar scientific objectives, researchers and engineers for both missions are working closely together. Moreover, since the costs of such missions are literally "astronomical," in the order of several billion dollars, NASA and the ESA may very well combine them.

And what will happen after Darwin? Most likely increasingly sensitive space-telescope arrays will be developed in the future. Proposals have already been made for multiple-element arrays that could reveal surface detail on Earth-like planets, including oceans and continents. Undoubtedly that will usher in an entirely new phase of planetary exploration and probably lead to discoveries we cannot even contemplate at this point. ■

Project OWL

Designed and planned by scientists and engineers at the European Southern Observatory (ESO), project OWL ranks as one of the most beautiful and ambitious scientific projects at the start of this century. This OverWhelmingly Large Telescope will incorporate many features of the VLT, which is currently the most powerful modern telescope. With its 100 m primary mirror, superb optics and novel technologies, the OWL telescope will be the largest, most powerful and most complex astronomical facility ever conceived.

The OWL project represents a radical departure from the 400-year tradition of telescope design, wherein astronomers and opticians would cautiously double the scale of existing instruments every few decades to build the largest telescopes of their time. The largest of the great refractors of the late 19th century boasted a 1 m objective lens, and in the 1920s the first giant reflector was installed at Mount Wilson Observatory in California with a 2.5 m mirror. This was doubled to 5 m by the 1950s with the installation of the 530-metric-ton Hale telescope at Mount Palomar. Finally in the 1990s, the twin 10 m Keck telescopes were built atop Hawaii's Mauna Kea volcano.

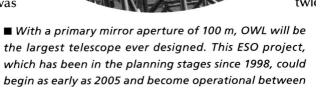

■ *With a primary mirror aperture of 100 m, OWL will be the largest telescope ever designed. This ESO project, which has been in the planning stages since 1998, could begin as early as 2005 and become operational between 2015 and 2020.*

OBSERVING THE BIRTH OF THE FIRST STARS

What motivated European astronomers to seriously consider a telescope like OWL, an instrument *ten* times larger than the biggest telescopes in the world today? Having just completed the ambitious VLT project, ESO scientists knew that modern telescopes had very little in common with the massive and very heavy instruments of the past. The largest telescopes today are comparatively lightweight optoelectronic machines and entirely computer controlled. To scale up such instruments entails little

more than building costlier receivers and support instrumentation and providing a lot more computing power. Even more important, constructing the primary mirror for such telescopes is no longer the technological challenge it used to be. That's because mirrors for 10 m class instruments and larger are no longer cast and polished as single units but are assembled as mosaics of smaller mirrors. This makes it possible in principle to build increasingly larger mirrors by simply adding additional segments. That is both technically and economically advantageous, since it is comparatively inexpensive to manufacture hundreds and even thousands of similar 1 to 2 m mirrors, something that has been done now for many years.

If the European community approves funding for the project, OWL might well become an international undertaking, much like the ALMA interferometer. According to ESO estimates, the project will cost less than US $1 billion, about three times as much as the VLT and twice the cost of ALMA. If the ESO and any potential partners approve the project by 2005, it would most likely be sited at an Atacama Desert location in Chile and become operational between 2015 and 2020.

Although the final optical configuration of the telescope must still be validated, its preliminary feasibility, cost and scientific merit assessment was completed in 2003. OWL will be a fully steerable, altazimuth-mounted instrument outwardly similar to the 100 m Green Bank radio telescope. Nearly 120 m tall, the telescope's enormous mount will have a moving mass of about 12,000 metric tons. Although OWL's optical configuration is intended to be as compact as possible, it will still be extraordinarily large and complex. The telescope's 100 m spherical primary mirror will be composed of 1,600 smaller 2.3 m diameter segments. The secondary mirror, which reflects light back to the focal plane, will also be huge and segmented, consisting of two hundred 2.3 m units arranged into a 34 m mosaic! OWL will be equipped with active optics designed to correct the primary mirror's aberrations and also to provide real time compensation for instrument flexure, air turbulence and wind buffeting. Consisting of two 8 m flexible mirrors, a 4.3 m passive mirror and an additional deformable 2.5 m mirror, this assembly will subsequently direct the light onto a system of adaptive optics that will deliver near-perfect images at both optical and infrared wavelengths.

OWL's optical performance should readily supersede anything attained by existing instruments. Its 100 m diameter mirror will attain resolution up to 0.001 arcsecs in visible light and record objects of 29th magnitude with exposures of a less than a minute. That is near the effective magnitude limit currently attained with the Hubble Space Telescope and the Keck, Gemini, Subaru and VLT facilities. OWL is expected to reach 35th magnitude with a one-hour exposure and its theoretical limit of 38th magnitude with exposures lasting several dozen hours.

ESO engineers are currently considering what types of imaging detectors could be scaled up for such a large telescope. Ideally astronomers would like both infrared and visible range detectors with a 1 x 1 degree field of view. This will require significant advances in technology, however, since to take full advantage of the telescope's planned 1 to 5 m focal ratio and exceptional resolving power, any camera based on CCD technology would have to be about 1 square meter and encompass several billion pixels to meet those requirements.

From a scientific standpoint, OWL's extraordinary 0.001 arcsec resolving power represents an unprecedented advance in optical astronomy. The telescope will able to directly image surface detail on thousands of nearby stars. It could monitor surface changes on Betelgeuse, for example, a red supergiant whose disk measures about 0.056 arcsecs. OWL will be able to image it with 2,500 pixel elements, sufficient to detect surface changes lasting just a few hours. The telescope will also show unprecedented detail in circumstellar disks like those of Beta Pictoris. Since this star is located only 60 light-years from us, OWL will resolve detail of only a few million kilometers in its gas disk. Should there be an asteroid belt or even small planets in this system, they should reveal themselves through gravitational perturbations in the dust disk, much like Saturn's moons affect its grooved ring system. Moreover, should any of Beta Pictoris' putative planets be close in size to Venus, Earth or Mars, OWL will probably detect them directly. The giant telescope will also directly image and spectroscopically analyze hundreds of exoplanets up to 500 light-years from us.

OWL's observational potential is equally daunting as we move further out in the cosmos. Thanks to its 100 m mirror, which will provide a

OWL'S ADAPTIVE OPTICS SYSTEM

Project OWL's success will depend entirely on the development of active and adaptive optics capable of delivering the high resolution and sensitivity of the instrument. It is also acknowledged that this will be the project's biggest technological challenge. First off, the system will require several thousand fast actuators on deformable mirrors to continuously compensate for wavefront distortions as starlight traverses the atmosphere. Existing active optics with just a few hundred actuators cannot simply be scaled up for this purpose; an entirely new system based on novel microtechnologies will have to be developed.

Another challenge is that OWL's adaptive optics will have to monitor several reference stars simultaneously across a much wider field of view than is currently possible. This will also require higher computing capabilities than presently available, but that is likely to change in the coming decade. Although these complex systems are not yet available, it is likely they can be developed in time and major efforts toward that end are currently underway in Europe and elsewhere.

hundredfold more light-gathering power than a 10 m telescope, OWL will also undertake detailed studies of almost all types of stars in nearby galaxies, including the Magellanic Clouds, NGC 6822 in Sagittarius, NGC 300 and 253 in Sculptor, and of course the well known Andromeda Galaxy and M33 in Triangulum. This will provide researchers detailed information on several billion stars and help clarify the evolution of other galaxies as well as our own. Paradoxically, OWL may at times be "blinded" by the brightest stars in our own and nearby galaxies, since stars brighter than 26th magnitude may well overwhelm its electronic sensors. In short, OWL may at times provide too much light-gathering power.

To further illustrate the immense light-gathering power of OWL, modest stars like our Sun will be within reach and amenable to detailed spectral analysis as far out as the Virgo galaxy cluster, 50 million light-years from the Milky Way. These galaxies will be fully resolved into myriads of individual stars.

OWL will also fully resolve bright stars like Vega, Arcturus and Aldebaran as far out as the Coma cluster, 300 million light-years from us.

THE TIME MACHINE

OWL will really come into its own observing the farthest reaches of the cosmos. Astronomers will be able to trace the evolution of the Universe all the way back to the birth of the first stars and primordial galaxies. The brightest known stars in the Universe are blue supergiants like Deneb and Rigel, with absolute magnitudes between −7 and −9. OWL will be powerful enough to potentially detect such stars even at redshifts around 2, or about 12 billion years ago when the Universe was only 75 percent its present age. Beyond that even a 100 m telescope cannot resolve individual stars, with the exception of supernovae. Such titanic

explosions, a billion times brighter than our Sun and lasting several weeks, tell us much about the chemical history of the Universe. They also serve as ideal cosmological markers. By obtaining very accurate apparent luminosity measurements of such objects at various distances from us and comparing them to luminosities of nearby supernovae, researchers hope to test relativistic properties resulting from the curvature of space/time. In theory, therefore, by observing a sufficient number of supernovae with OWL, cosmologists hope to have enough information to fully define the expansion rates of space/time, record any changes over time, and attain a fuller understanding of the exact age and shape of the Universe.

OWL might also be able to look back into the so-called "Dark Ages" of the early Universe. This timeframe is currently inaccessible to us, since any objects that distant have redshifts greater than 6, extending back 14 billion years to when the first primordial galaxies were formed. Finally, OWL might also detect the supernova signatures of the very first stars in the cosmos. It is still unclear how far back those processes extend: probably beyond redshifts of 15 and more, less than 100 million years after the big bang. OWL might well provide astronomers with enough information to trace the entire history of our universe, from the very first stars and galaxies to the present. That might also bring us to the limit of observational cosmology. The mysteries of the big bang itself would be left solely to physicists, being outside the realm of observational astronomy. It is more likely, however, that human curiosity will lead us in entirely unforeseen directions, with new questions to ask and problems to solve. Who knows, by then astronomers may well be planning time machines more powerful than OWL to probe even deeper into the mysteries of the cosmos. ■

The Space Hyper-Telescope and Exo-Earth Imager

I t's probably the "ultimate" telescope. The kind of instrument you dream about, one that literally lets you visit other worlds, as if traveling in some future starship. While it is highly unlikely that actual interstellar travel will be possible any time soon, French astronomer and physicist Antoine Labeyrie has proposed a telescope that could actually image Earth-size planets around nearby stars. Called the Exo-Earth Imager (EEI) hyper-telescope, such an instrument could be built this century and launch us on a *virtual* voyage through the Universe. An ambitious and costly project like this would only make sense to scientists and taxpayers if they felt the investment was worth it. For that to happen, prior space missions like COROT, Eddington, SIM, Darwin and TPF would first have to find Earth-like planets around other stars, and then show that they have atmospheres and temperatures similar to ours and contain water or at least water vapor: in short that they might actually harbor life.

I N T E R N A T I O N A L

■ *This artist's rendition shows what a space hyper-telescope consisting of one hundred and fifty 3 m telescopes might look like. An instrument like this could become reality 30 years from now.*

Only then is a project of the scale of EEI likely to be undertaken, since only giant space interferometers would let us observe such planets in detail.

This is where the hyper-telescope proposed by Antoine Labeyrie comes into play. Astronomers and optical designers have generally assumed that it was not possible to scale up an imaging interferometer beyond a certain size. To retain optimal imaging capabilities, the ratio between the actual collecting area (the sum of all optical surfaces in the array) and the simulated area (the maximum distance between the individual telescopes in the array) of such an instrument has to be kept within reasonable limits and

avoid aperture "dilution." For example, the Large Binocular Telescope at Mount Graham Observatory combines two separate 8.2 m mirrors, giving it the resolving power of a simulated 22.8 m telescope but an effective aperture of only 12 m. In this situation, the optical dilution is only 3.7 and the instrument delivers high-quality images over a wide field and very short exposure times. On the other hand, despite having four 8.2 m mirrors, the VLTI telescopes are distributed across more than 3 hectares, resulting in a 150-fold optical dilution. Because of this, the array requires long image integration times and various baseline combinations and is limited to observing rather bright objects in a very small field of view. Thus despite its impressive 0.001 arcsec resolution, the VLTI will never be able to produce complex images with subtle brightness and contrast variations similar to what its four telescopes are capable of individually.

Labeyrie's innovative hyper-telescope design would use "stippled telescopes," as he puts it. It would deploy a very large number of small aperture telescopes to simulate an enormous aperture interferometer that would be larger and more homogenous than any other design. The array would also include a novel technique called "density pupil imaging," producing good images of faint distant planets even with significant aperture dilution. The French astronomer indicates that after spacecraft like Darwin and TPF have identified possible Earth-type extrasolar planets, the EEI hyper-telescope could be deployed to reveal actual surface detail on them.

The EEI would be an array of perhaps one hundred and fifty

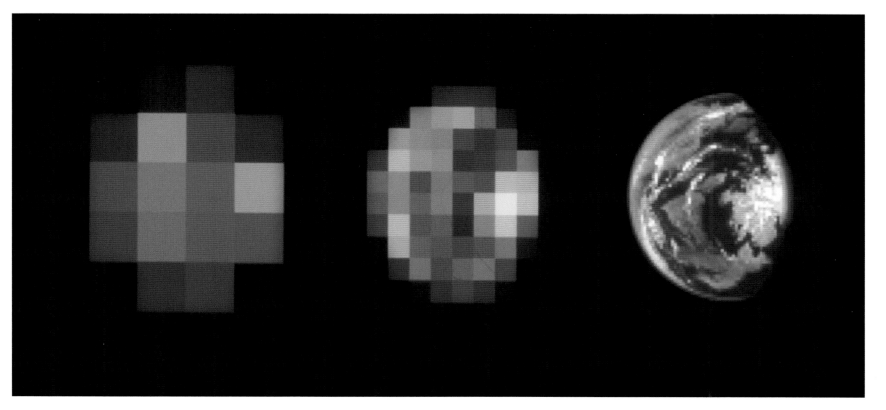

■ *These three images simulate an Earth-like planet located 30 light-years from us. From left to right: surface detail is shown as it would appear with space hyper-telescopes of 75, 150 and 1,500 m effective aperture.*

3 m aperture space telescopes, flying in formation to simulate a hyper-telescope with an effective aperture of about 150 km. Despite the modest apertures of each individual telescope, the full array would have the light-gathering capability of a single 36 m mirror. Working at both visible and infrared wavelengths, the array would be positioned and controlled by laser beams to an accuracy of a few nanometers, similar to the Michelson interferometer design used with the gravity wave interferometer (LISA). With its 150 km aperture, the EEI should be able to image 30th to 32nd magnitude planets with exposures of just a few minutes, and resolve detail as small as 0.000,001 or 1 micro arcsecs!

The capabilities of such an instrument would be truly astounding. It could reveal surface detail on stars several thousand light-years away and observe the dust and gas envelopes circling the massive black hole at the Milky Way's core. At this distance, about 28,000 light-years, it could pick out structures as small as one million km and objects that have been gravitationally distorted by the black hole itself. The EEI's primary mission, however, would be to study and map exoplanets. Since it should resolve detail less than 1,000 km across on planets within a 20 light-year radius from us, it could discern atmospheric patterns, continents and mountain chains, large impact craters, oceans and polar ice caps on such worlds. Seasonal changes could be monitored and direct evidence of life might be found in the form of large expanses of vegetation similar to the Amazon basin. In fact, the EEI would be able to scrutinize planets around nearby stars to the same level of detail as the Hubble Space Telescope currently does with the more distant planets and satellites of the Solar System.

The EEI would actually let us explore other planetary systems telescopically, much like 19th and 20th century observers did within our solar system. The difference would be that EEI could reach much further and probably encompass thousands of new planets as far out as 200 to 400 light-years. Moreover, such planets would be studied in far greater detail. In addition to such basics as rotation periods, surface and (when present) atmospheric temperatures, EEI would provide detailed information of the chemical makeup and composition of exoplanets and any large moons orbiting them. Most importantly, however, the hyper-telescope would be able to tell us whether any such planets are habitable or perhaps even inhabited.

The cost of such a project would itself be astronomical, and estimated at around US $12 billion. While clearly extremely expensive, the EEI would still cost less than the Apollo program, the International Space Station or the estimated cost of a manned mission to Mars. Perhaps then the EEI price tag is not that great, considering it would open up a virtual exploration of many other worlds.

Theoretically, at least, there is no real limit on the size a hyper-telescope of this type could be. If 100 years from now interstellar travel continues to be impossible, perhaps even larger instruments of this sort will be built. Large enough perhaps to directly study life forms on other distant Earths. ■

Appendices

■ Located more than 400 light-years away, the spiral galaxy IGC 10214 was recorded in this eight-hour exposure by the Hubble Space Telescope. The faintest objects in this image are 29th magnitude. In the background are hundreds of galaxies, many several billion light-years away.

NORTH

AMERICA

Maple Ridge
Victoria

Hanford

Sudbury

Mont Mégantic

David Dunlap
Yerkes

Hat Creek

Owens Valley
Mount Hamilton
Anderson Mesa
Mount Graham
VLBA
Big Bear
SOR
Mount Wilson
VLA — Sac Peak
Mount Palomar
Kitt Peak
Apache Point
San Pedro Martir
Mount Locke
Cananea
Mount Hopkins

Green Bank

Livingston

Haleakala
Mauna Kea
Mauna Loa

Sierra Negra

Arecibo

ATLANTIC

OCEAN

AFRICA

PACIFIC

OCEAN

Kvistaberg
Ondrejov — Onsala
Kanzelhöhe
Mont Ekar
Tautenburg
Westerbork
Effelsberg
Jodrell Bank
Cambridge
Merlin
Gornergrat
Nançay
Plateau de Bure
Haute-Provence
Pic du Midi
Calern
Junchfraujoch
Calar Alto
Pico Veleta
Wendelstein
Virgo
Medicina
Gran Sasso
Catania

Pico de Teide
La Palma

Merida

SOUTH

AMERICA

Chacaltaya

Cerro Chajnantor
Cerro Paranal

Pico dos Dias

Las Campanas
La Silla
Cerro Pachon
Cerro Tololo
Leoncito
Pierre Auger

South Pole Observatory

160

120

80

40

0

observatories

80°

Arctic Circle

Tartu

UROPE

n

naté Pleso

t Lomnicky

t Piszkéstetö

Crimée

Zelentchouk

Zelentchouk

Byurakan

ASIA

Miyun • Xinglong

40°

Maidanak

Mont Chelmos

Kamioka

Mount Bohyun

Nobeyama

Okayama

Norikura

tamia

Mount Saraswati

Tropic of Cancer

Narayangaon

PACIFIC

OCEAN

Kavalur

INDIAN

Equator

OCEAN

Bras d'Eau

Tropic of Capricorn

AUSTRALIA

erland

Narrabri

Siding Spring

Parkes

40°

ANTARCTICA

Concordia Station

40   80   120   160

Antarctic Circle

Pic du Midi
Observatory

Haute-Provence
Observatory

Calern
Observatory

Plateau de Bure
Observatory

Calar Alto
Observatory

Pico de Teide
Observatory

La Palma
Observatory

Jodrell Bank Radio
Observatory

Merlin
Interferometer

The 100 largest

The Pic du Midi Observatory
42°56' N, 0°08' E
Altitude: 2,870 m
Instruments: 2, 1, .6 and .5 m telescopes;
a .5 m solar refractor and a .25 m
coronagraph
www.obs-mip.fr/omp/Pic/index.html
France

The Haute-Provence Observatory
43°56' N, 5°42' E
Altitude: 650 m
Instruments: 2, 1.5, 1.2 and .8 m telescopes
www.obs-hp.fr/
France

The Calern Observatory
43°45' N, 6°55' E
Altitude: 1,270 m
Instruments: an interferometer with two
1.5 m telescopes; a 1.5 m lunar-laser
telescope; a laser-satellite telescope; a
.9 m Schmidt telescope, a heliometer
and a gamma-burst detector
www.obs-azur.fr/
France

The Plateau de Bure Observatory
44°38' N, 5°54' E
Altitude: 2,550 m
Instrument: a millimeter and submillimeter
interferometer with six 15 m antennae
www.iram.fr/
France

The Nançay Radio Telescope
47°22' N, 2°11' E
Altitude: 150 m
Instruments: a 200 x 35 m meridian radio
telescope, and a radio heliograph
www.obs-nancay.fr/
France

The Pico Veleta Antenna
37°04' N, 3°24' W
Altitude: 2,870 m
Instrument: a 30 m millimeter wavelength
radio telescope
www.iram.fr/
Spain

The Calar Alto Observatory
37°14' N, 2°32' W
Altitude: 2,160 m
Instruments: 3.5, 2.2, 1.5, 1.2, and .8 m
telescopes
www.mpia-hd.mpg.de/Public/index_en.html
Spain

The Pico de Teide Observatory
28°18' N, 16°30' W
Altitude: 2,400 m
Instruments: .9, .6, .04 and .4 m solar
telescopes; 1.5, 1 and .8 m telescopes and
millimeter detectors
www.iac.es/ot/indice.html
Spain

La Palma Observatory
28°45' N, 17°53' W
Altitude: 2,426 m
Instruments: 10, 4.2, 3.6, 2.55, 2,
1 and .6 m telescopes,
and cosmic-ray detectors
www.ing.iac.es/
Spain

The Jodrell Bank Radio Observatory
53°14' N, 2°18' W
Altitude: 78 m
Instruments: 76 and 25 m radio telescopes
www.jb.man.ac.uk/
United Kingdom

The Merlin Interferometric Array
53°14' N, 2°18' W
Altitude: 78 m
Instrument: an interferometer with
seven dishes ranging between
76, 32 and 25 m
www.merlin.ac.uk/
United Kingdom

The Cambridge Interferometer
52°09' N, 0°03' W
Altitude: 17 m
Instrument: an interferometer consisting
of five .4 m radio telescopes
www.mrao.cam.ac.uk/telescopes/coast/
United Kingdom

The Tartu Observatory
58°16' N, 26°28' E
Altitude: 10 m
Instruments: 1.5, .6 and .5 m telescopes
www.aai.ee/
Estonia

The Kvistaberg Observatory
59°30' N, 17°36' E
Altitude: 34 m
Instruments: a 1 m Schmidt telescope
and a .4 m telescope
www.astro.uu.se/history/Kvistaberg.html
Sweden

The Onsala Observatory
57°24' N, 11°55' E
Altitude: 24 m
Instruments: 25 and 20 m millimeter
radio telescopes
www.oso.chalmers.se/
Sweden

The Westerbork Interferometer
52°55' N, 6°36' E
Altitude: 16 m
Instrument: a decimeter wavelength
interferometer consisting of fourteen
25 m antennae
www.astron.nl/p/observing.htm
The Netherlands

The Gornergrat Observatory
45°59' N, 7°47' E
Altitude: 3,135 m
Instruments: a 3 m submillimeter
wavelength telescope,
a 1.5 m infrared telescope
www.ph1.uni-koeln.de/gg
Switzerland

The Jungfraujoch Observatory
46°33' N, 7°59' E
Altitude: 3,580 m
Instruments: heliometers and
cosmic-ray detectors
www.ifjungo.ch/
Switzerland

The Effelsberg Radio Telescope
50°31' N, 6°53' E
Altitude: 320 m
Instrument: a 100 m radio telescope
www.mpifr-bonn.mpg.de/div/effelsberg/
index_e.html
Germany

The VLBI-JIVE array
Longitude: from 70°W, to 120°E
Instruments: a centimeter radio
interferometer, consisting
of 18 dishes measuring 14, 20,
25, 32 and 100 m
www.evlbi.org/evn.html
Europe

The Wendelstein Observatory
47°42' N, 12°01' E
Altitude: 1,845 m
Instruments: an .8 m telescope and a
.2 m coronograph
www.usm.uni-muenchen.de/people/
hbarwig/wst/wst_en.html
Germany

The Tautenburg Observatory
50°58' N, 11°42' E
Altitude: 342 m
Instrument: a 2 m telescope
www.tls-tautenburg.de/
Germany

The Ondrejov Observatory
49°54' N, 14°47' E
Altitude: 534 m
Instruments: a 2 m telescope, a .25 m
spectroheliograph and a radio telescope
sunkl.asu.cas.cz/~radio/
The Czech Republic

The Skalnaté Pleso Observatory
49°11 N, 20°14' E
Altitude: 1,782 m
Instruments: .6 m telescopes and a
very large widefield camera
www.astro.sk/homepage.php
Slovakia

Effelsberg Radio
Telescope

Virgo Gravity
Wave Telescope

Zelentchouk
Observatory

Arecibo Radio
Telescope

Green Bank
Radio Telescope

Very Large Array

Very Long
Baseline Array

Starfire
Optical Range

Sacramento
Peak
Observatory

world observatories

The Mount Lomnicky Station
49°11' N, 20°13' E
Altitude: 2,632 m
Instrument: a .2 m double coronograph
Slovakia

The Kanzelhöhe Observatory
46°40' N, 13°54' E
Altitude: 1,526 m
Instruments: solar .4 and .2 m telescopes
www.solobskh.ac.at/
Austria

The Mount Piszkéstetö Observatory
47°55' N, 19°53' E
Altitude: 958 m
Instruments: 1, .6 and .5 m telescopes,
and a .6 m Schmidt telescope
www.konkoly.hu/konkoly/konkoly.html
Hungary

The Virgo Gravity Wave Telescope
43°37' N, 10°30' E
Altitude: 5 m
Instruments: A gravity-wave interferometer
www.virgo.infn.it/
Italy

The Gran Sasso Observatories
42°37' N, 13°30' E
Altitude: 960 and 2,188 m
Instruments: neutrino detectors
and 1.1 and .9 m telescopes
www.lngs.infn.it/
Italy

The Medicina Radio Telescope
44°31' N, 11°38' E
Altitude: 42 m
Instrument: a 560 m x 35 m radio
interferometer, a 32 m antenna as
part of the VLBI
www.ira.cnr.it/
Italy

The Mount Ekar Observatory
45°50' N, 11°34' E
Altitude: 1,350 m
Instruments: 1.8 and .9 m telescopes
www.pd.astro.it/asiago/
Italy

The Catania Observatory
37°41' N, 14°58' E
Altitude: 1,735 m
Instruments: .9, .8 and .6 m telescopes,
and a .2 m solar refractor
webusers.ct.astro.it/ccdlab/cold/
Italy

The Mount Chelmos Observatory
38°00' N, 22°21' E
Altitude: 2,340 m
Instrument: a 2.3 m telescope
www.astro.noa.gr/ASC_2.3m/ngt_site.htm
Greece

The Torun Observatory
53°05' N, 18°33' E
Altitude: 80 m
Instruments: the 32 and 15 m VLBI
array antennae
www.astro.uni.torun.pl/
Poland

The Zelentchouk Observatory
43°39' N, 41°26' E
Altitude: 2,070 m
Instruments: 6 and 1 m telescopes
www.sao.ru/Doc-en/index.html
Russia

The Zelentchouk Radio Telescope
43°49' N, 41°35 E
Altitude: 970 m
Instrument: a 600 m circular
radio telescope
www.sao.ru/Doc-en/Telescopes/ratan/
descrip.html
Russia

The Byurakan Observatory
40°20' N, 44°18' E
Altitude: 1,450 m
Instruments: a 2.6 m telescope and
a 1 m Schmidt telescope
Armenia

The Crimean Observatory
44°43' N, 34°00' E
Altitude: 600 m
Instruments: 2.6, 1.26, 1.25, and 1.2 m
telescopes
www.crao.crimea.ua/
The Ukraine

The Dominion Astrophysical Observatory
48°31' N, 123°25' W
Altitude: 240 m
Instruments: 1.8, 1.2 and .4 m telescopes
www.hia-iha.nrc-cnrc.gc.ca/index.html
Canada

The UBC Liquid-Mirror Observatory
49°10' N, 122°32' W
Altitude: 395 m
Instrument: a 6 m liquid mirror
zenith telescope
www.astro.ubc.ca/LMT/lzt/index.html
Canada

David Dunlap Observatory
43°51' N, 79°25' W
Altitude: 395 m
Instrument: a 1.9 m telescope
www.ddo.astro.utoronto.ca/ddohome/
Canada

The Mont Mégantic Observatory
45°27' N, 71°09' W
Altitude: 1,114 m
Instruments: 1.6 and .6 m telescopes
www.astro.umontreal.ca/omm/index.html
Canada

The Sudbury Neutrino Observatory
46°30' N, 81°00' W
Altitude: 1,780 m below sea level
Instruments: neutrino detectors
www.sno.phy.queensu.ca/
Canada

The Arecibo Radio Telescope
18°20' N, 66°45' W
Altitude: 496 m
Instrument: a 300 m radio telescope
www.naic.edu/about/ao/descrip.htm
Puerto Rico, United States

The Hanford Gravitational Telescope
46°30' N, 119°20' W
Altitude: 180 m
Instrument: a gravitational wave detector
www.ligo-wa.caltech.edu/#welcome
Washington State, United States

The Livingston Gravitational Telescope
30°40' N, 90°50' W
Altitude: 120 m
Instrument: a gravitational wave detector
www.ligo-la.caltech.edu/public.htm
Louisiana, United States

The Green Bank Radio Observatory
38°25' N, 79°50' W
Altitude: 836 m
Instruments: 100, 42, 25 and
20 m radio telescopes
www.gb.nrao.edu/GBT/GBT.html
West Virginia, United States

The Yerkes Observatory
42°34' N, 88°33' W
Altitude: 334 m
Instruments: a 1 m refractor; 1 and .6 m
telescopes
astro.uchicago.edu/yerkes/
Wisconsin, United States

The Very Large Array
34°04' N, 107°37' W
Altitude: 2,200 m
Instrument: an interferometer array of
twenty-seven 25 m radio telescopes
www.vla.nrao.edu/
New Mexico, United States

Apache Point
Observatory

Kitt Peak
Observatory

Mount Hopkins
Observatory

Mount Graham
International
Observatory

Mount Locke
Observatory

Mount Wilson
Observatory

Haleakala
Observatory

Mauna Kea
Observatory

Cerro Tololo
Observatory

The Very Long Baseline Array
Longitude: from 65°to 155°W
Instrument: an interferometer consisting
of ten 25 m antennae
www.aoc.nrao.edu/vlba/html/VLBA.html
United States

The Starfire Optical Range
34°58' N, 106°28' W
Altitude: 1,900 m
Instruments: 3.5 and 1.5 m telescopes
New Mexico, United States

The Sacramento Peak Observatory
32°47' N, 105°49' W
Altitude: 2,800 m
Instruments: a .76 m solar telescope;
a .4 m coronagraph
www.sunspot.noao.edu/index.html
New Mexico, United States

The Apache Point Observatory
32°47' N, 105°49' W
Altitude: 2,800 m
Instruments: 3.5, 2.5 and 1 m telescopes
www.apo.nmsu.edu/
New Mexico, United States

The Anderson Mesa Observatory
35°05' N, 111°32' W
Altitude: 2,200 m
Instruments: 1.8, 1, .8 m telescopes;
an infrared interferometer consisting of
six .5 m telescopes
www.lowell.edu/
Arizona, United States

The Kitt Peak Observatory
31°58' N, 111°36' W
Altitude: 2,060 m
Instruments: 3.8, 3.5, 2.4, 2.3, 2.1, 1.3 and
.9 m telescopes; a 1.6 m solar telescope
and a 25 m VLBA antenna
www.astro.lsa.umich.edu/obs/mdm/
Arizona, United States

The Mount Hopkins Observatory
31°41' N, 110°53' W
Altitude: 2,600 m
Instruments: 6.5, 1.5,
1.3 and 1.2 m telescopes; an infrared
interferometer with three .45 m telescopes
and a cosmic-ray detector
cfa-www.harvard.edu/mmt/
Arizona, United States

The Mount Graham International Observatory
32°42' N, 109°53' W
Altitude: 3,170 m
Instruments: a 2 x 8.4 m telescope;
a 1.8 m telescope and a 10 m
submillimeter radio telescope
mgpc3.as.arizona.edu/
Arizona, United States

The Mount Locke Observatory
30°40' N, 104°01' W
Altitude: 2,072 m
Instruments: 9.2, 2.7, 2.1 and .8 m
telescopes
www.as.utexas.edu/mcdonald/het/het.html
Texas, United States

The Hat Creek Observatory
40°49' N, 121°28' W
Altitude: 1,044 m
Instrument: a millimeter interferometer
consisting of ten 6 m antennae
bima.astro.umd.edu/
California, United States

The Mount Wilson Observatory
34°13' N, 118°03' W
Altitude: 1,742 m
Instruments: 2.5 and 1.5 m telescopes; an
optical interferometer consisting of six 1 m
telescopes and an infrared interferometer
consisting of two 1.6 m telescopes
www.mtwilson.edu
California, United States

The Mount Palomar Observatory
33°21' N, 116°52' W
Altitude: 1,706 m
Instruments: 5 and 1.5 m telescopes,
a 1.2 m Schmidt telescope and an infrared
interferometer consisting of three .5 m
telescopes
www.astro.caltech.edu/palomar/
California, United States

The Mount Hamilton Observatory
37°20' N, 121°38' W
Altitude: 1,290 m
Instruments: 3 and 1 m telescopes and an
.89 m refractor
www.irving.org/xplore/lick/contents.html
California, United States

The Big Bear Solar Observatory
34°15' N, 116°55' W
Altitude: 2,067 m
Instrument: a solar telescope made up
of .65, .25, .2 and .15 m optics
www.bbso.njit.edu/
California, United States

The Haleakala Observatory
20°42' N, 156°15' W
Altitude: 3,050 m
Instruments: 3.7, 1.6, 1.2 m and .8 m
telescopes and .5 and .25 m coronagraph
www.ifa.hawaii.edu/haleakala/
Hawaii, United States

The Mauna Loa Observatory
19°32' N, 155°34' W
Altitude: 3,398 m
Instruments: a .23 m coronograph, a
spectroheliograph and a solar polarimeter
www.mlo.noaa.gov/default.htm
Hawaii, United States

The Mauna Kea Observatory
19°49' N, 155°28' W
Altitude: 4,208 m
Instruments: Two 10 m telescopes;
8.3, 8.1, 3.8, 3.6, 3, 2.2 and .6 m
telescopes; 15 and 10.4 m millimeter
radio telescopes; a millimeter
interferometer consisting of eight 6 m
antennae and a 25 m VLBA antenna
www.ifa.hawaii.edu/mko/
Hawaii, United States

The Sierra Negra Observatory
18°59' N, 97°30' W
Altitude: 4,640 m
Instrument: a 50 m millimeter wavelength
radio telescope
www.lmtgtm.org
Mexico

The Cananea Observatory
31°03' N, 110°23' W
Altitude: 2,480 m
Instruments: 2.1 and .4 m telescopes
www.inaoep.mx/~astrofi/cananea/
Mexico

The San Pedro Martir Observatory
31°02' N, 115°27' W
Altitude: 2,480 m
Instruments: 2.1, 1.5 and .85 m telescopes
www.astrosen.unam.mx/
Mexico

The Llano del Hato Observatory
08°47' N, 70°52' W
Altitude: 3,610 m
Instruments: a 1 m telescope, a
1 m Schmidt telescope and .65 and
.5 m refractors
www.cida.ve/
Venezuela

The Pico dos Dias Observatory
22°32' S, 45°35' W
Altitude: 1,864 m
Instruments: 1.6 m, two .6 m telescopes
www.lna.br/Welcom_e.html
Brazil

The Leoncito Observatory
31°47' S, 69°18' W
Altitude: 2,552 m
Instrument: a 2.1 m telescope
www.casleo.gov.ar/
Argentina

The Pierre Auger Observatory
35°00' S, 69°00' W
Altitude: 1,500 m
Instrument: a cosmic-ray detector
www.auger.org/
Argentina

The Chacaltaya Observatory
16°21' S, 68°08' W
Altitude: 5,220 m
Instruments: cosmic-ray detectors
Bolivia

Cerro Pachon
Observatory

Las Campanas
Observatory

La Silla
Observatory

Cerro Paranal
Observatory

Siding Spring
Observatory

The Parkes Radio
Observatory

The Narrabri
Interferometer

The Narayangaon
Interferometer

Kamioka Neutrino
Observatory

The Cerro Tololo Observatory
30°10' S, 70°49' W
Altitude: 2,200 m
Instruments: 4, 1.5, 1.3, 1, .9 and
.6 m telescopes
www.ctio.noao.edu/
Chile

The Cerro Pachon Observatory
30°21' S, 70°49' W
Altitude: 2,715 m
Instruments: 8.1 and 4.2 m telescopes
www.gemini.edu
Chile

Las Campanas Observatory
29°00' S, 70°42' W
Altitude: 2,300 m
Instruments: two 6.5 m telescopes;
2.5 m, 1 and .3 m telescopes
www.ociw.edu/lco/
Chile

La Silla Observatory
29°15' S, 70°44' W
Altitude: 2,400 m
Instruments: 3.6, 3.5, 2.2, 1.54, 1.2
and 1 m telescopes; a 15 m millimeter
wavelength antenna
www.ls.eso.org/index.html
Chile

The Cerro Paranal Observatory
24°37' S, 70°24' W
Altitude: 2,635 m
Instruments: Four 8.2 m telescopes; a
2.5 m telescope; three 1.8 m telescopes
www.eso.org/paranal/
Chile

The Cerro Chajnantor Observatory
23°00' S, 67°45' W
Altitude: 5,050 m
Instruments: a compact millimeter
interferometer; a millimeter wavelength
interferometer array of sixty-four 12 m
dishes
www.eso.org/projects/alma
Chile

Kottamia Observatory
29°56' N, 31°49' E
Altitude: 475 m
Instrument: a 1.9 m telescope
Egypt

Sutherland Observatory
32°22' S, 20°48' E
Altitude: 1,798 m
Instruments: 9.2, 1.9, 1, .75 and .5 m
telescopes
www.salt.ac.za/
South Africa

The Bras d'Eau Observatory
20°10' S, 57°40' E
Altitude: 20 m
Instrument: a 2,048 x 880 m
wavelength radio telescope
icarus.uom.ac.mu/mrt2.html
Mauritius

The Siding Spring Observatory
31°16' S, 149°04' E
Altitude: 1,150 m
Instruments: 3.9, 2.3 and 1.2 m telescopes
www.aao.gov.au/
Australia

The Parkes Radio Observatory
33°00' S, 148°15' E
Altitude: 392 m
Instrument: a 76 m radio telescope
www.parkes.atnf.csiro.au/
Australia

The Narrabri Interferometer
30°19' S, 149°33' E
Altitude: 218 m
Instrument: a radio interferometer
consisting of six 22 m dishes
www.narrabri.atnf.csiro.au/
Australia

The Mount Maidanak Observatory
38°41' N, 66°56' E
Altitude: 2,700 m
Instruments: 1.5 and 1 m telescopes and
solar instruments
www.academy.uz/eng/objekts/astrin.html
Uzbekistan

The Narayangaon Interferometer
19°06' N, 74°03' E
Altitude: 650 m
Instrument: a radio interferometer
consisting of thirty 45 m dishes
India

The Mount Saraswati Observatory
32°46' N, 78°58' E
Altitude: 4,500 m
Instruments: 2 and .5 m telescopes
www.iiap.res.in/iao/site.html
India

The Kavalur Observatory
12°34' N, 78°50' E
Altitude: 700 m
Instruments: 2.3, 1 and .75 m telescopes
www.vellore.tn.nic.in/kavalur.htm
India

The Xinglong Observatory
40°23' N, 117°34' E
Altitude: 960 m
Instruments: 4.2, 2.1, 1.2, .9, .85 and
.6 m telescopes
China

The Miyun Observatory
40°34' N, 116°46' E
Altitude: 160 m
Instrument: a meter wavelength
interferometer consisting of twenty-eight
9 m antennae
www.bao.ac.cn/bao/index-e.html
China

The Mount Bohyun Observatory
36°09' N, 128°58' E
Altitude: 1,124 m
Instruments: a 1.8 m telescope and
a .5 m solar telescope
www.boao.re.kr/
South Korea

The Kamioka Observatory
36°25' N, 137°18' E
Altitude: 350 m
Instruments: neutrino detectors
www-sk.icrr.u-tokyo.ac.jp/
Japan

The Nobeyama Observatory
35°56' N, 138°29' E
Altitude: 1,350 m
Instruments: a 45 m millimeter antenna,
and a millimeter interferometer consisting
of seven 10 m dishes
www.nro.nao.ac.jp/index-e.html
Japan

The Norikura Observatory
36°07' N, 137°33' E
Altitude: 2,876 m
Instruments: a .25 m coronagraph
and two .1 m coronagraphs
solarwww.mtk.nao.ac.jp/english/
norikura.html
Japan

The Okayama Observatory
34°34' N, 133°35' E
Altitude: 372 m
Instruments: 1.9 and .9 m telescopes;
a .65 m solar telescope
www.cc.nao.ac.jp/oao/index.html
Japan

The South Pole Observatory
90°00' S, 0°00'
Altitude: 2,840 m
Instruments: a neutrino detector; a
.6 m infrared telescope; 1.7 and 2 m
submillimeter telescopes
www.nsf.gov/od/opp/antarct/treaty/
projsum01/html/cara.html
and
amanda.berkeley.edu/amanda/
amanda.html
Antarctica

The Concordia Station
75°00' S, 123°0' E
Altitude: 3,250 m
Instruments: an array of solar, infrared and
millimeter wavelength telescopes
Antarctica

The 50 largest astronomical telescope mirrors

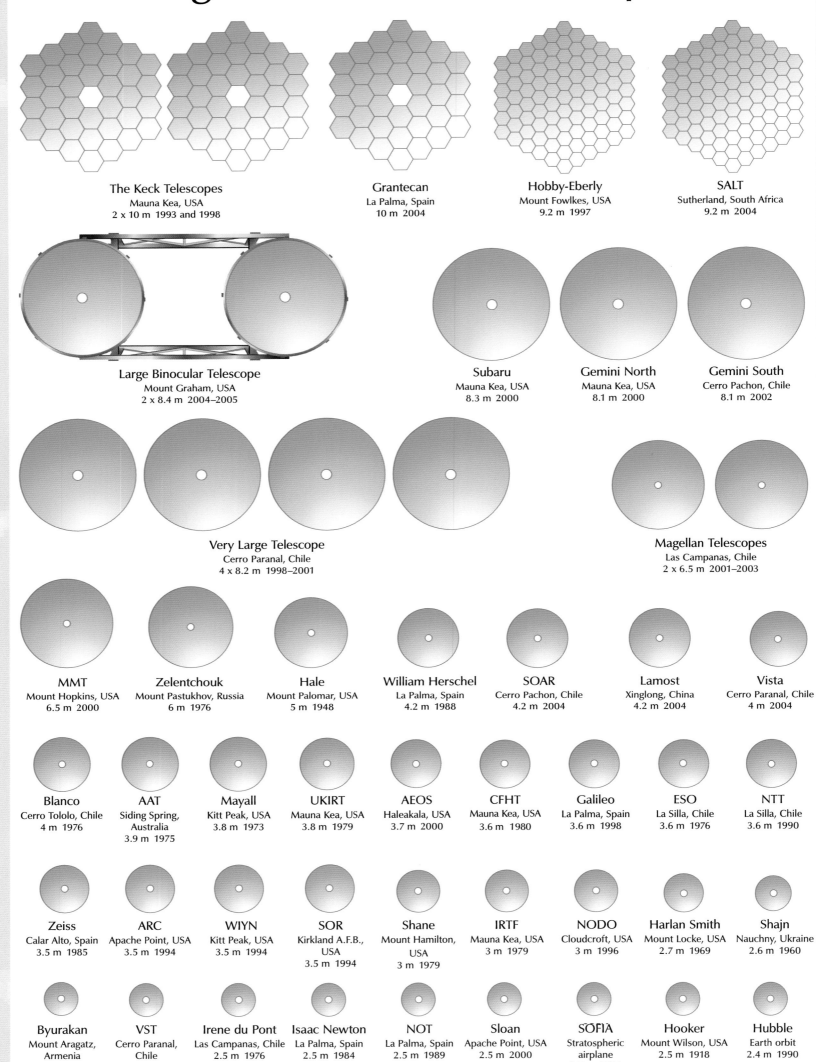

The Keck Telescopes
Mauna Kea, USA
2 x 10 m 1993 and 1998

Grantecan
La Palma, Spain
10 m 2004

Hobby-Eberly
Mount Fowlkes, USA
9.2 m 1997

SALT
Sutherland, South Africa
9.2 m 2004

Large Binocular Telescope
Mount Graham, USA
2 x 8.4 m 2004–2005

Subaru
Mauna Kea, USA
8.3 m 2000

Gemini North
Mauna Kea, USA
8.1 m 2000

Gemini South
Cerro Pachon, Chile
8.1 m 2002

Very Large Telescope
Cerro Paranal, Chile
4 x 8.2 m 1998–2001

Magellan Telescopes
Las Campanas, Chile
2 x 6.5 m 2001–2003

MMT
Mount Hopkins, USA
6.5 m 2000

Zelentchouk
Mount Pastukhov, Russia
6 m 1976

Hale
Mount Palomar, USA
5 m 1948

William Herschel
La Palma, Spain
4.2 m 1988

SOAR
Cerro Pachon, Chile
4.2 m 2004

Lamost
Xinglong, China
4.2 m 2004

Vista
Cerro Paranal, Chile
4 m 2004

Blanco
Cerro Tololo, Chile
4 m 1976

AAT
Siding Spring,
Australia
3.9 m 1975

Mayall
Kitt Peak, USA
3.8 m 1973

UKIRT
Mauna Kea, USA
3.8 m 1979

AEOS
Haleakala, USA
3.7 m 2000

CFHT
Mauna Kea, USA
3.6 m 1980

Galileo
La Palma, Spain
3.6 m 1998

ESO
La Silla, Chile
3.6 m 1976

NTT
La Silla, Chile
3.6 m 1990

Zeiss
Calar Alto, Spain
3.5 m 1985

ARC
Apache Point, USA
3.5 m 1994

WIYN
Kitt Peak, USA
3.5 m 1994

SOR
Kirkland A.F.B.,
USA
3.5 m 1994

Shane
Mount Hamilton,
USA
3 m 1979

IRTF
Mauna Kea, USA
3 m 1979

NODO
Cloudcroft, USA
3 m 1996

Harlan Smith
Mount Locke, USA
2.7 m 1969

Shajn
Nauchny, Ukraine
2.6 m 1960

Byurakan
Mount Aragatz,
Armenia
2.6 m 1976

VST
Cerro Paranal,
Chile
2.5 m 2003

Irene du Pont
Las Campanas, Chile
2.5 m 1976

Isaac Newton
La Palma, Spain
2.5 m 1984

NOT
La Palma, Spain
2.5 m 1989

Sloan
Apache Point, USA
2.5 m 2000

SOFIA
Stratospheric
airplane
2.5 m 2003

Hooker
Mount Wilson, USA
2.5 m 1918

Hubble
Earth orbit
2.4 m 1990

Field of view, magnitude and resolution

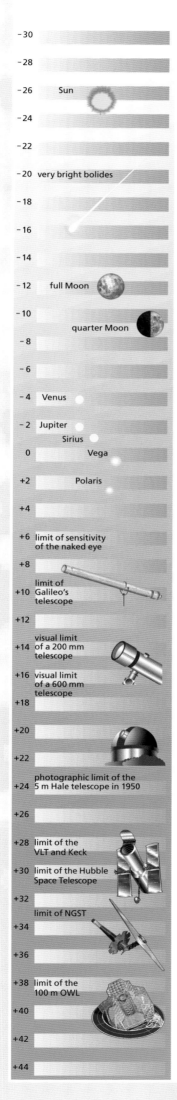

Astronomers always consider three basic criteria when specifying performance characteristics of telescopes: field of view, resolving power and collecting power. The field of view is an angular measure expressed in degrees (°), arc minutes (') and arc seconds ("), where 1°= 60' = 3,600". To clarify this: the celestial sphere covers 41,244 square degrees; the horizon extends in a 360° circle; the distance from horizon to zenith is 90°; and the largest field covered by current electronic cameras today is 1°.

Telescopic resolving power, or the ability to see fine detail, is also an angular measure. The naked eye cannot discern detail smaller than about 60 arcsec on the Moon, which measures about 1/2°, which is 30 arcmin or 1,800 arcsec. Galileo resolved about 10 arcsec with his small refractor, and large telescopes cannot resolve binaries closer than 0.5 to 1 arcsec because of atmospheric turbulence. The largest telescopes in the world today, like the two Kecks, Gemini, Subaru and the VLT, can reach about 0.05 arcsec resolution with adaptive optics, about the same range attainable with the Hubble Space Telescope. Fairly soon, the best optical interferometers should reach 0.001 arcsec resolution.

The collecting capacity of a telescope is its ability to detect radiation from faint or weak sources. For more than 2,000 years astronomers have classified celestial objects according to their apparent brightness or magnitude. One magnitude difference between two stars corresponds to a brightness factor of 2.51. For example, a 5th magnitude object is a hundred times fainter than a 0 magnitude star. The Sun's magnitude is –26.7, while the 20 brightest stars in the sky all fall roughly between magnitude –1 and +1.

The faintest stars visible by the naked eye are around 6th magnitude, and a small beginner's telescope will reach 10th magnitude. A 4 m telescope can register stars to 26th magnitude with a one-hour exposure: a hundred million times fainter than the dimmest object detectable by the naked eye. The most powerful modern telescopes, like Hubble and VLT, have limiting magnitudes between 29 and 30. The difference in brightness between the Sun and the faintest star reached by Hubble exceeds 56 magnitudes, or about one hundred trillion.

■ The apparent stellar magnitude scale is open-ended. The brightest objects, like the Sun and Moon, have negative values. The brightness difference between magnitude values across the scale is 2.51, or 100 times for 5 magnitudes, 10,000 times for 10.

■ The four images above show the galaxy M51 in Canes Venatici imaged by instruments with different fields of view, resolution and light-gathering power. At magnitude 9, this object is not visible to the naked eye and is barely visible in the top photograph taken with a modest telephoto lens.

In the second image, the galaxy's spiral structure is quite evident in a 45-minute exposure taken with a small amateur refracting telescope.

In the third image, a 30-minute exposure with the 3.6 m CFHT reveals many of the galaxy's individual gas nebulae. This 0.8 arcsec resolution image shows stars down to about 26th magnitude.

In the bottom image, a six-hour exposure with the Hubble Space Telescope and its very small 3 x 2 arcmin field shows extremely fine detail (0.1 arcsec) in the galaxy's nucleus. Several individual 26th magnitude stars are also fully resolved.

The History of Light

It took many centuries of observation and experimenting before the physical nature of light was fully clarified. At first the properties of light could only be described in empirical terms, but over time increasingly sophisticated theoretical models were developed to explain them. These developments were often tied to advances in observational astronomy.

5th century B.C.—According to Heraclitus of Ephesus, light is the fire that unites all things. For many centuries after this, there is more interest in vision than in the nature of light itself.

3rd century B.C.—Euclid introduces the idea that light rays travel in straight lines.

2nd century—Galen of Pergamus investigates the structure of the eye.

11th century—The Arab physicist Ibn al-Hayth, better known as Alhazen, does the first experiments showing how sunlight interacts with the eye. By reviving the idea of light rays, he is the first to link "object" and "image," and the first to describe such effects as reflection, refraction and dispersion of light. He also shows that light rays could be crossed without being affected.

15th century—Leonardo da Vinci studies the propagation of light and is interested in the similarities between light and sound.

1621—Englishman Willebrord Snell formulates the laws of refraction.

1637—Rene Descartes outlines the geometric basis of optics.

1665—Pierre de Fermat declares that light travels between two points in the shortest possible time, a principle that also provides a theoretical basis for the laws of reflection and refraction.

1665—Robert Hooke suggests that light is propagated through vibration.

1665—Christian Huygens proposes the first wave theory of light, with waves traveling through a medium termed the ether.

1685—Father Grimaldi discovers diffraction.

1672—Isaac Newton enunciates the corpuscular theory of light.

1675—Expanding on his theory, Newton suggests that waves of ether guide particles of light.

1676—Olaüs Römer determines that the speed of light is less than 350,000 km/s by observing occultations of Jupiter's satellites.

1704—Isaac Newton proposes his theory of color after studying the prismatic dispersion of white light.

1760—Pierre Bouguer studies photometry.

1800—William Herschel proposes that "heat" rays from the Sun behave in a similar manner to rays of visible light. These were later termed "infrared" rays.

1808—Etienne Malus shows that light is polarized.

1816—Thomas Young explains the phenomenon of interference using the wave theory of light.

1819—Augustin Fresnel and François Arago undertake experiments that support the wave theory of light.

1889—François Arago discovers the principles of the daguerreotype, precursor to photography.

1842—Christian Doppler shows that frequency shifts arise when sources are moving (Doppler effect).

1849—Hippolyte Fizeau undertakes the first direct measurement of the speed of light.

1850—Léon Foucault measures the speed of light in various media, including water. He shows that light travels more slowly in water than in air, supporting the wave theory of light.

1859—Gustav Kirchhoff and Robert Bunsen develop spectral analysis.

1871—Emulsions are invented for photographic plates.

1873—James Clerk Maxwell proposes his electromagnetic theory and the equations governing it.

1879—Joseph Stefan defines the laws of black-body radiation.

1881—Albert Michelson shows experimentally that the speed of light is independent of the point of reference from which it is measured.

1888—Heinrich Hertz generates electromagnetic waves using oscillating electric sparks.

1896—Wilhelm Roentgen discovers X-rays.

1896—Pieter Zeeman shows that light rays are affected by strong magnetic fields.

1905—Albert Einstein outlines the theory of relativity, revives the particulate nature of light and introduces the concept of photons.

1916—Albert Einstein proposes the notion of gravitational waves in the context of the theory of general relativity.

1920—Louis de Broglie finally reconciles the particulate and wave nature of light.

1932—Wolfgang Pauli predicts the existence of neutrinos.

1950—Hannes Alfven discovers magnetohydrodynamic waves.

HISTORY OF ASTRONOMICAL INSTRUMENTS AND METHODS

1590—The optical telescope is invented.

1610—Galileo Galilei uses his famous telescope and among other things discovers Jupiter's four principal satellites.

1670—Isaac Newton builds the first reflecting telescope and a few years later develops the first transit telescope.

1781—William Herschel discovers Uranus using a 1.2 m telescope.

1814—Joseph von Fraunhofer observes the solar spectr and discovers dark absorption lines that he cannot explain.

1845—First daguerreotype image of the Sun.

1870—William Parsons builds a telescope equipped with a 1.82 m speculum mirror.

1871—John W. Rayleigh explains the blue color of the sky as light scattering (Rayleigh scattering).

1872—Henry Draper obtains the first photographic image of a stellar spectr, that of Vega.

1878—Construction of Pic du Midi observatory begins in France.

1880—Henry Draper takes the first photograph of a nebula.

1891—George Ellery Hale invents the spectroheliograph and analyzes the solar atmosphere.

1920—Albert Michelson uses interferometry to obtain the first measurements of stellar diameters.

1932—Karl Jansky detects radio waves from our galaxy. This marks the beginning of radioastronomy.

1936—André Lallemand invents the electronic camera.

1945—Haute-Provence Observatory is established in France.

1946—The first ultraviolet observations of the Sun are obtained.

1948—The 5 m telescope is inaugurated at Mount Palomar (United States).

1954—The radio observatory at Nançay (France) is established.

1957—Ultraviolet radiation from a star is detected for the first time.

1961—Frank Low invents the first bolometers.

1962—The European Southern Observatory (ESO) is established.

1963—The first celestial X-ray observations are made.

1963—The 300 m Arecibo radio telescope enters service.

1965—Arno Penzias and Robert Wilson discover cosmic microwave background radiation.

1970—The Westerbork radio interferometer enters service in the Netherlands.

1972—The first gamma-ray signals are detected.

1972—The 100 m Effelsberg radio telescope enters service.

1975—The COS-B gamma-ray telescope is launched into orbit.

1979—The Canada–France–Hawaii Telescope enters service.

1979—The Einstein X-ray telescope is launched.

1980—The VLA radio interferometer enters service.

1983—The IRAS infrared space telescope is launched.

1985—The 30 m IRAM radio telescope enters service at Pico Veleta.

1989—The Hipparcus spacecraft is launched.

1990—The Compton and Hubble space telescopes are launched.

1990—The first system of adaptive optics is used at La Silla in Chile.

1990—The IRAM radio interferometer is established at Bure plateau.

1993—The 10 m Keck I telescope enters service, followed by Keck II in 1996.

1998—The 8 m telescope of the VLT enters service, followed by the other three at one-year intervals.

1999—The 8 m Gemini North telescope and the 8 m Subaru enter service.

1999—The Newton X-Ray MM Space Telescope is launched.

2001—Gemini South is inaugurated.

2001—The Keck and VLT interferometers see first light.

2001—The INTEGRAL gamma-ray space telescope is launched.

2002—The Spitzer Space Telescope is launched (sub-mm/IR).

2004—The Einstein Gravity Probe B is launched.

Constants and fundamental units

As in all branches of physics, astronomical measurements and quantities have little meaning unless they are expressed in specific units, such as wavelengths, energy or time. However, given the nature and scale of the different objects of interest to scientists, many of the internationally accepted units are not always convenient to apply in practice. For example, it is not practical to express either the extremely small size of atoms or the huge dimensions of our galaxy in meters. Instead, appropriate units and measures must be applied in each case.

Fundamental constants have also been introduced that apply to specific laws of physics: for example, in electricity, magneti and gravity. These too have to be defined in specific units.

INTERNATIONAL UNITS AND DERIVATIONS

Basic units like meters, kilograms and seconds represent key reference points in the physical sciences. Established by international agreement, these units form part of the International System of Units (SI), which is widely used to define all other units. Other, units, called derived units, have been introduced to deal with energy, electricity and other important physical parameters; these too are expressed in units defining length, mass, time, charge, temperature, etc.

SI UNITS

length: meter (m)
mass: kilogram (kg)
time: second (s)
electric charge: coulomb (C)
temperature: Kelvin (K)

SOME DERIVED UNITS

force: newton (N) $1N = 1kg.m/s^2$
energy: joule (J) $1J = 1N.m$
power: watt (W) $1W = 1J/s$
electric current: ampere (A) $1A = C/s$
potential: volts (V) $1V = 1J/C$
frequency: hertz (Hz) $1Hz = 1/s$
pressure: pascal (Pa) $1Pa = 1N/m^2$

SOME USEFUL CONSTANTS

c (speed of light) $= 2.9979 \times 10^8$ m/s
e (electron charge) $= 1.6022 \times 10^{-19}$ C
h (Planck's constant) $= 6.6262 \times 10^{-34}$ Js
k (Boltzmann's constant) $= 1.3807 \times 10^{-23}$ J/K
G (gravitational constant) $= 6.6726$ N m²/kg²
$\mu 0$ (permeability of free space)
$= 4\pi \times 10^{-7}$ N/A²
0 (permitivity of free space)
$= 8.8542 \times 10^{12}$ C²/N.m²
s (Stefan–Boltzmann's constant)
$= 5.6703 \times 10^{-8}$ W/m².K⁴

UNITS OF LENGTH BASED ON THE METER

1 angstrom $= 10^{-10}$ m

1 nanometer $= 10^{-9}$ m

1 micron $= 10^{-6}$ m

1 astronomical unit (AU) = average distance between Earth and Sun = 149.5979 million km

1 light-year = distance traveled by light in one year $= 9.4605 \times 10^{15}$ m

1 parsec $= 3.085810^{16}$ m = 3.2616 light-years = 206,265 AU

UNITS OF TIME BASED ON THE SECOND

1 hour = 3,600 s

1 sidereal day = 86164.10 s

1 sidereal year $= 3.156 \times 10^7$ s

ANGULAR UNITS

1 arcsec = 0.0000048481 radian

1 arcmin = 60 seconds = 0.0002908882 radian

1 degree = 60 minutes = 0.017453925 radian

1 radian = 57.2957795141 degrees

1 hour = 15 degrees

1 minute = 15 arcmin

1 second of time = 15 arcsec

UNITS OF ENERGY AND FLUX—PHOTOMETRY

ENERGY

1 erg $= 10^{-7}$ J

1 calorie = 4.1854 J

1 electron volt (eV) $= 1.6 \times 10^{-19}$ J $= 1.6 \times 10^{-12}$ erg. The energy acquired by an electron when accelerated by a potential of 1 Volt.

Energy of a photon of wavelength
$(\mu m) + 1.98648 \times 10^{-12}/\lambda(erg)$

POWER

1 watt = 1 J/s

Blackbody radiation: The brightness of a black object; this can be expressed in terms of wavelength (or monochromatic radiation) or as frequency. Brightness expressed in terms of wavelength is the intensity of energy emitted per surface area, per unit angle and per unit of wavelength:

$B_\lambda = 2hc^2/\lambda^5.(e^{h.c/k\lambda T}-1)$ W.m⁻²m⁻¹ster⁻¹
$= 1.191 \times 10^{-16}/\lambda^5.(e^{0.01439/\lambda T}-1)$ W.m⁻²m⁻¹ster⁻¹

Brightness expressed as frequency is the intensity of radiation emitted by an object per surface area, per unit angle and per unit of frequency:

$B_\lambda = 2h\nu^5/c^2.(e^{h.\lambda/kT}-1)$ W.m⁻²m⁻¹ster⁻¹

If $h.\nu << kT$, the expressions may be reduced to an approximation known as the Rayleigh-Jeans equation:

$B_\lambda = 2ck$ T/λ⁴ W.m⁻²m⁻¹ster⁻¹
$= 8.2782 \times 10^3/\lambda_{\mu m}^4$ W.m⁻²μm⁻¹ster⁻¹
$B_\nu = 2k.T.\nu^2/c^2$ W.m⁻²Hz⁻¹ster⁻¹
$= 3.0724 \times 10^{-40}$ T.ν^2 W.m⁻²Hz⁻¹ster⁻¹

BRIGHTNESS AND FLUX

Flux is the power of an electric field striking a surface perpendicular to the direction of propagation. In other words, it is the power per unit wavelength in the spectral range under consideration. This is usually expressed in W/m².μm. Brightness is a function of wavelength or the spectral domain of the radiation in question and is expressed as power per unit wavelength in W/m². Astronomers often use the term "flux" incorrectly when referring to brightness; the correct unit for flux is the Jansky and is equivalent to 10^{-26}W.m⁻²Hz⁻¹

MAGNITUDES

The apparent magnitude of a point source (or star) at a given wavelength λ is defined by the following expression: m_λ= constant − 2.5 log (e_λ), where e_λ designates the brightness of the object at that wavelength as measured from Earth and the constant defines the magnitude scale. By convention this system sets Vega at 0 magnitude at all wavelengths. The following relationship hold trues therefore for two objects of apparent magnitude, m¹ and m², respectively: m¹-m² = 2.5log (e²/e¹), where e¹ and e² represent the brightness of each respective object at a given wavelength λ.

Stellar magnitudes can also be expressed as an integrated value within a defined spectral domain. For instance, magnitudes may be measured within the spectral bands U, B, V, R, I, J, H, K, L, M, N, Q. When all these values are combined across the full spectr, they are termed bolometric magnitude.

Lastly, we must also mention the absolute magnitude scale (designated by an upper case, M). This corresponds to the brightness of an object measured from a distance of 10 parsecs. The following relationship holds: M = m+5−5.log(D)−A, where M is the absolute magnitude, D its distance and A its interstellar absorption value.

Bibliography

Recommended references and readings

Joseph Ashbrook, *Astronomical Scrapbook*, Sky Publishing, 1984.

Jean Audouze and Guy Israel, *The Cambridge Atlas of Astronomy*, Cambridge University Press, 1993.

Louis Bell, *The Telescope*, Dover, 1981.

T. Dickinson, *The Universe and Beyond*, Firefly Books, 2004.

R. Guenther, *Modern Optics*, John Wiley and Sons, 1990.

Henry C. King, *History of the Telescope*, Dover, 1980.

C.R. Kitchin, *Astrophysical Techniques*, Adam Hilger, 1991.

John Lankford, *History of Astronomy: An Encyclopedia*, Garland Publishing, 1996.

Richard Learner, *Fenêtres sur l'Univers*, Denoël, 1981.

I.S. MacLean, *Electronic Imaging in Astronomy*, John Wiley and Sons, 1997.

P. Murdin and Margaret Penston, *The Firefly Encyclopedia of Astronomy*, Firefly Books, 2004.

Thornton Page, *Telescopes*, MacMillan, 1966.

Harrie Rutten and Martin van Venrooij, *Telescope Optics*, Willmann-Bell, 1988.

Peter Shaver, *Science with Large Millimeter Arrays*, Springer, 1996.

Robert W. Smith, *The Space Telescope*, Cambridge, 1989.

Henry Stockman, *The Next Generation Space Telescope*, AURA, 1997.

Raymond Wilson, *Reflecting Telescope Optics I: Basic Design Theory and Its Historical Development*, Springer, 1996.

Raymond Wilson, *Reflecting Telescope Optics II: Manufacture, Testing, Alignment, Modern Techniques*, Springer, 1999.

Selected List of Astronomical and Space Web Sites

The French magazine *Ciel & Espace*:
http://www.cieletespace.fr/

Sky & Telescope:
http://www.skypub.com/

Space News:
http://www.space.com

The National Aeronautics and Space Administration:
http://www.nasa.gov/

The European Space Agency (ESA):
http://www.esa.int/

The French space agency,
National Center for Space Studies (CNES):
http://www.cnes.fr/

The European Space Agency science site:
http://sci.esa.int/

All space science missions:
http://spacescience.nasa.gov/missions/index.htm

The Darwin exoplanet space mission:
http://ast.star.rl.ac.uk/darwin/mission.html

Terrestrial Planet Finder mission:
http://planetquest.jpl.nasa.gov/TPF/tpf_index.html

The COROT exoplanet space mission:
http://www.obspm.fr/encycl/corot.html

The Hubble Space Telescope:
http://hubblesite.org

Direct access to SOHO solar images:
http://sohowww.nascom.nasa.gov/cgi-bin/
soho_images_bydate?summary

The future James Webb Space Telescope:
http://www.jwst.nasa.gov/

The Jean Schneider Extrasolar Planet Encyclopedia:
http://cfa-www.harvard.edu/planets/f-encycl.html

Bill Arnett's listing of the largest telescopes in the world:
http://www.seds.org/billa/bigeyes.html

The European 100 m OWL project telescope:
http://www.eso.org/projects/owl/index_3.html

The Californian 30 m CELT telescope project:
http://celt.ucolick.org/

The American 30 m GSMT telescope project:
http://staging.noao.edu/gsmt.html

The European 50 m Euro-50 telescope project:
http://www.astro.lu.se/~torben/euro50/index.html

Index

Photo Credits

Front inside and back cover, S. Brunier/*Ciel et Espace*; **1** S. Brunier/*Ciel et Espace*; **3** S. Brunier/*Ciel et Espace*; **5** and copy p. 137, ESO; **8** Museum of Science, Florence, Ph Scala; **9** National library, Florence, Ph Scala; **10** TOP LEFT National Library of France, Paris, Ph AKG Paris; **10** BOTTOM RIGHT Ph Coll. Archives Larbor; **11** TOP RIGHT Paris Observatory; **11** BOTTOM LEFT Dr. J. Burgess/S.P.L./Cosmos; **12** TOP LEFT Explorer Archives; **12** BOTTOM RIGHT Mary Evans/Explorer Archives; **13** TOP RIGHT J. Burgess/S.P.L./Cosmos; **13** BOTTOM S.P.L./Cosmos; **14** TOP LEFT AKG Paris; **14** BOTTOM RIGHT Paris Observatory; **15** TOP RIGHT coll. part, AKG Paris; **15** BOTTOM LEFT Lowell Observatory Archives; **16** TOP LEFT Mount Wilson and Palomar Observatories; **16** BOTTOM RIGHT Huntington Library and Art Gallery; **17** courtesy of the Archives, California Institute of Technology; **18** and **19,** J.C. Cuillandre/CFHT/*Ciel et Espace*; **23** TOP S. Brunier/*Ciel et Espace*; **23** BOTTOM S. Brunier/*Ciel et Espace*; **24** and copy p. 228, S. Brunier/*Ciel et Espace*; **23** TOP S. Brunier/*Ciel et Espace*; **23** BOTTOM S. Brunier/*Ciel et Espace*; **24** and copy p. 228, S. Brunier/*Ciel et Espace*; **25** C. Voulgaropoulos/*Ciel et Espace*; **26** BOTTOM NOAO/AURA/NSF; **26** BOTTOM RIGHT Trace Data Center/LMSAL; **27** L. Blondel; **28** TOP ESA; **28** MIDDLE NOAO/AURA/NSF; **28** BOTTOM Atlas Image courtesy of 2MASS/UMass/IPAC-Caltech/NASA/NSF; **29** ESO; **30** and copy p. 228, S. Brunier/*Ciel et Espace*; **31** S. Brunier/*Ciel et Espace*; **32** S. Brunier/*Ciel et Espace*; **33** S. Brunier/*Ciel et Espace*; **34** S. Brunier/*Ciel et Espace*; **35** and copy p. 228, J. Sandford/S.P.L./Cosmos; **36** and copy p. 228, P. Henarejos/*Ciel et Espace*; **37** I. Marquez/J.A. Bonet/Astrophysical Institute of the Canaries; **38** ESO; **39** L. Blondel; **40** TOP D. Parker/S.P.L./Cosmos; **40** BOTTOM D. Parker/S.P.L./Cosmos; **41** TOP Courtesy Neelon Crawford/Polar Fine Arts/Gemini Observatory; **41** BOTTOM ESO; **42** Royal Greenwich Observatory/S.P.L./Cosmos; **43** and copy p. 228, M. Van Der Hoeven/Isaac Newton Group of Telescopes, La Palma; **44** and **45,** 8 photos courtesy of the Isaac Newton Group of Telescopes, La Palma; **46** and copy p. 228, S. Brunier/*Ciel et Espace*; **47** TOP S. Brunier/*Ciel et Espace*; **47** BOTTOM D. Parker/S.P.L./Cosmos; **48** and copy p. 228, S. Brunier/*Ciel et Espace*; **49** 3 photos Merlin/Jodrell Bank Observatory; **50** NRAO/*Ciel et Espace*; **51** TOP AND BOTT, 2 photos S. Brunier/*Ciel et Espace*; **51** MIDDLE A. Cirou/*Ciel et Espace*; **52** P. Boulat/Cosmos; **53** and copy p. 229, M. Bond/S.P.L./Cosmos; **54** photo Plailly/Eurelios; **55** TOP and copy p. 229, photo Plailly/Eurelios; **55** BOTTOM photo Plailly/Eurelios; **56** S. Brunier/*Ciel et Espace*; **57** and copy p. 229, S. Brunier/*Ciel et Espace*; **58** P. Parviainen/*Ciel et Espace*; **59** TOP L. Blondel; **59** BOTTOM ESO; **60** and copy p. 229, C. Lebedinsky/*Ciel et Espace*; **61** S. Brunier/*Ciel et Espace*; **62** TOP and p. 63 TOP, C. Lebedinsky/*Ciel et Espace*; **62** BOTTOM S. Brunier/*Ciel et Espace*; **63** BOTTOM C. Lebedinsky/*Ciel et Espace*; **64** M. Bailey/NRAO/AUI; **65** and copy p. 229, M. Bailey/NRAO/AUI; **66** S. Brunier/*Ciel et Espace*; **67** and copy p. 229, S. Brunier/*Ciel et Espace*; **68** TOP and p. 69 TOP, S. Brunier/*Ciel et Espace*; **68** BOTT and p. 69 BOTTOM, 2 photos NRAO/*Ciel et Espace*; **70** L. Blondel; **71** NOAO; **72** TOP NOAO; **72** BOTTOM L. Blondel/NASA; **73** TOP AND BOTT, 2 photos ESO; **74** S. Brunier/*Ciel et Espace*; **75** TOP LEFT and copy p. 229, R. Ressmeyer/Corbis; **75** TOP RIGHT and BOTTOM LEFT and RIGHT, 3 photos S. Brunier/*Ciel et Espace*; **76** S. Brunier/*Ciel et Espace*; **77** M. Rupen et al./NRAO; **78** J.A. Sugar/Corbis; **79** TOP and copy p. 230, R. Ressmeyer/Corbis; **79** BOTTOM C. Gino/Mount Wilson Institute; **80** D. Parker/S.P.L./Cosmos; **81** and copy p. 203, S. Brunier/*Ciel et Espace*; **82** TOP and p. 83 TOP, D. Parker/S.P.L./Cosmos; **82** BOTTOM M. Pierce/J. Jurcevic/WIYN/NOAO/NSF; **83** BOTTOM C. Howk/B. Savage/N. Sharp/WIYN/NOAO/NSF; **84** J.C. Cuillandre/CFHT/1998; **85** ESO; **86** STScI/NASA; **87** ESO; **88** and copy p. 229, S. Brunier/*Ciel et Espace*; **89** S. Brunier/*Ciel et Espace*; **90** and copy p. 230, Fermilab/S.P.L./Cosmos; **91** Stephen Kent, SDSS Collaboration; **92** R. Ressmeyer/Corbis; **93** TOP AND BOTT, and copy p. 230, 2 photos Howard Lester/The MMT Observatory, a joint facility of the Smithsonian Institution and the University of Arizona; **94** CEA/DAPNIA; **95** ESO; **96** TOP S. Brunier/*Ciel et Espace*; **96** BOTTOM CEA/DAPNIA; **97** J.C. Cuillandre/CFHT/*Ciel et Espace*; **98** S. Brunier/*Ciel et Espace*; **99** and copy p. 229, S. Brunier/*Ciel et Espace*; **100** and copy p. 230, S. Criswell; **101** Large Binocular Telescope Project; **102** D. Parker/S.P.L./Cosmos; **103** TOP and copy p. 230, McDonald Observatory; **103** BOTTOM McDonald Observatory; **104** S. Brunier/*Ciel et Espace*; **105** ESO; **106** TOP AND BOTT, 2 photos ESO; **107** TOP LEFT Courtesy of the Navy Prototype Optical Interferometer at Lowell, a project of the U.S. Naval Observatory, The Naval Observatory in association with Lowell Observatory; **107** TOP RIGHT ESO; **108** and copy p. 230, S. Brunier/*Ciel et Espace*; **109** TOP S. Brunier/*Ciel et Espace*; **109** BOTTOM J.L. Heudier/Cerga/*Ciel et Espace*; **110** S. Brunier/*Ciel et Espace*; **111** S. Brunier/*Ciel et Espace*; **112** © 1999–2001 Subaru Telescope, NAOJ. All rights reserved; **113** S. Brunier/*Ciel et Espace*; **114** and copy p. 230, S. Brunier/*Ciel et Espace*; **115** © 1999–2001 Subaru Telescope, NAOJ. All rights reserved; **116** and p. 117, S. Brunier/*Ciel et Espace*; **118** and copy p. 230, NOAO/S.P.L./Cosmos; **119** NOAO/AURA/NSF; **120** ESO; **121** NOAO; **122** TOP ESO; **122** BOTTOM LEFT L. Blondel; **122** BOTTOM RIGHT ESO; **123** Keck Observatory/NASA; **124** courtesy Neelon Crawdord/Polar Fine Arts/Gemini Observatory; **125** and copy p. 231, courtesy Gemini Observatory; **126** and copy p. 231, Carnegie Observatories and the Carnegie Institution of Washington, D.C.; **127** LEFT and RIGHT, 2 photos B. Penprase, W. Freedman, B. Madore/Magellan, Carnegie Observatories and the Carnegie Institution of Washington, D.C.; **128** and copy p. 231, S. Brunier/*Ciel et Espace*; **129** ESO; **130** ESO; **131** ESO; **132** ESO; **133** ESO; **134** S. Brunier/*Ciel et Espace*; **135** TOP LEFT NASA/ESA; **135** TOP RIGHT A.M. Lagrange, D. Mouillet/Adonis/ESO; **136** ESO; **138** ESO; **139** and copy p. 231, ESO; **140** TOP AND BOTTOM, 2 photos ESO; **141** ESO; **142** ESO; **143** ESO; **144** and copy p. 231, S. Brunier/*Ciel et Espace*; **145** S. Brunier/*Ciel et Espace*; **146** Anglo-Australian Observatory and photograph by S. Lee, C. Tinney, and D. Malin; **147** S. Brunier/*Ciel et Espace*; **148** and copy p. 231, S. Brunier/*Ciel et Espace*; **149** S. Brunier/*Ciel et Espace*; **150** TOP RIGHT NASA/STScI; **150** BOTTOM LEFT L. Blondel; **151** TOP J.C. Cuillandre/CFHT/*Ciel et Espace*; **151** BOTTOM NASA/STScI; **152** and copy p. 231 S. Brunier/*Ciel et Espace*; **153** TOP AND BOTT, 2 photos S. Brunier/*Ciel et Espace*; **154** and p. 155, and copy p. 231, 2 photos B. Premkumar/NCRA-TIFR, Pune, India; **156** and p. 157, and copy p. 231, 3 photos ICRR (Institute for Cosmic Ray Research), The University of Tokyo; **158** and p. 159, ACS/NASA/ESA/STScI; **160** NASA; **161** NASA; **162** STScI/NASA; **163** NASA; **164** STScI/NASA; **165** STScI/NASA; **166** ESA; **167** TOP AND MIDDLE, 2 photos NASA; **168** ESA; **169** TOP AND BOTT, 2 photos ESA; **170** NASA/*Ciel et Espace*; **171** TOP NASA/UMass/D. Wang et al.; **171** BOTTOM NASA/CXC/U.Md/A. Wilson et al.; **172** NASA; **173** TOP TRW Inc.; **173** BOTTOM NASA/F. Baganoff et al.; **174** NASA; **175** TOP NASA/WMAP Science Team; **175** BOTTOM NASA; **176** FUSE/JHU/NASA; **177** ESO; **178** S. Brunier/*Ciel et Espace*; **179** S. Brunier/*Ciel et Espace*; **180** Courtesy of NASA/JPL/Caltech; **181** Courtesy of NASA/JPL/Caltech; **182** David Burrows, Penn State University; **183** ESO; **184** Artist's rendition of INTEGRAL, D. Ducros/ESA; **185** TOP integration of INTEGRAL at the ESTEC sites, A. Van Der Geest/ESA; **185** BOTTOM ESA; **186** Gravity Probe B, Stanford University; **187** Gravity Probe B, Stanford University; **188** and p. 189; D. Wittman/CTIO/NOAO; **190** L. Blondel; **191** TOP NRAO; **191** BOTTOM ESO; **192** and p. 193, illustration L. Blondel, based on an ESO image; **194** S. Brunier/S. Aubin/*Ciel et Espace*; **195** L. Blondel; **196** ESO; **197** ESO; **198** L. Blondel; **199** TOP AND BOTTOM, 2 photos ESA/ISO/ISOCAM/Alain Agbergel; **200** L. Blondel; **201** TOP ESA; **201** BOTTOM L. Blondel; **202** L. Blondel; **203** NASA/ESA/STScI/*Ciel et Espace*; **204** LEFT L. Blondel/ESO; **204** RIGHT ESO; **205** P. Barthel/M. Neeser/ESO; **206** MIDDLE AND BOTT, 4 photos J.C. Cuillandre/Canada–France–Hawaii Telescope/1999; **207** STScI/NASA; **208** L. Blondel; **209** NASA/ESA/STScI; **210** L. Blondel; **211** J. Ludriguss/*Ciel et Espace*; **212** NASA/ESA/STScI; **213** L. Blondel; **215** TOP J.C. Cuillandre/CFHT/*Ciel et Espace*; **215** BOTTOM N. Metcalfe, T. Shanks, University of Durham/WHT/ING; **216** NASA/ESA/STScI; **217** L. Blondel; **217** L. Blondel; **218** L. Blondel; **219** L. Blondel; **220** and p. 221, L. Blondel; **222** L. Blondel; **223** O. Hodasava/*Ciel et Espace*; **224** and p. 225, NASA/ESA/STScI; **226** and p. 227, L. Blondel; **232** L. Blondel; **233** LEFT L. Blondel; **233** TOP RIGHT A. Fujii/*Ciel et Espace*; **233** TOP MIDDLE RIGHT, J. Lodriguss/*Ciel et Espace*; **233** BOTTOM MIDDLE RIGHT, CFHT/J.C Cuillandre/*Ciel et Espace*; **233** NASA/STScI/NOAO/*Ciel et Espace*